W9-AHA-932

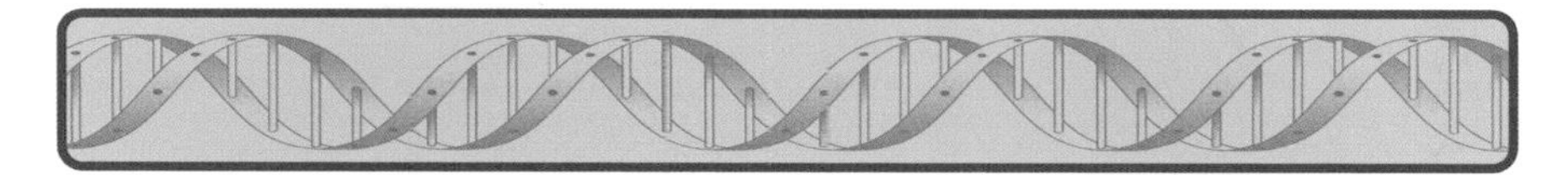

Biotechnology in the 21st Century

Bioinformatics, Genomics, and Proteomics

Getting the Big Picture

Biotechnology in the 21st Century

Biotechnology on the Farm and in the Factory
Agricultural and Industrial Applications

Biotechnology and Your Health
Pharmaceutical Applications

Bioinformatics, Genomics, and Proteomics
Getting the Big Picture

The Ethics of Biotechnology

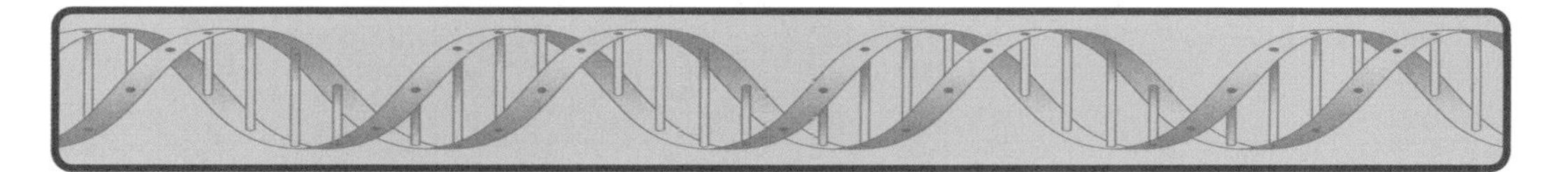

Biotechnology in the 21st Century

Bioinformatics, Genomics, and Proteomics

Getting the Big Picture

Ann Finney Batiza, Ph.D.

The author thanks Dr. Jean-Yves Sgro of the University of Wisconsin, Madison, for creating the molecular models shown in Figures 2.1 and 7.4, for proofreading the manuscript, and for many helpful comments. I also appreciate the calm hand of my editor, Beth Reger, the support of my children Eric and Rodey, and the patience of my husband, John Short, during the preparation of this book.

Bioinformatics, Genomics, and Proteomics

Chelsea House
An imprint of Infobase Publishing
132 West 31st Street
New York NY 10001

ISBN-10: 0-7910-8517-1
ISBN-13: 978-0-7910-8517-2

Library of Congress Cataloging-in-Publication Data
Batiza, Ann.
Bioinformatics, genomics, and proteomics: getting the big picture/Ann Batiza.
p. cm.
Includes bibliographical references and index.
ISBN 0-7910-8517-1 (alk. paper)
1. Bioinformatics—Popular works. 2. Genomics—Popular works. 3. Proteomics—Popular works.
I. Title.
QH324.2B38 2005
572.8'0285—dc22 2005017232

Chelsea House books are available at special discounts when purchased in bulk quantities for businesses, associations, institutions, or sales promotions. Please call our Special Sales Department in New York at (212) 967-8800 or (800) 322-8755.

You can find Chelsea House on the World Wide Web at http://www.chelseahouse.com

Text and cover design by Keith Trego

Printed in the United States of America

Bang 21C 10 9 8 7 6 5 4 3 2

This book is printed on acid-free paper.

All links and web addresses were checked and verified to be correct at the time of publication. Because of the dynamic nature of the web, some addresses and links may have changed since publication and may no longer be valid.

Table of Contents

Detailed Table of Contents

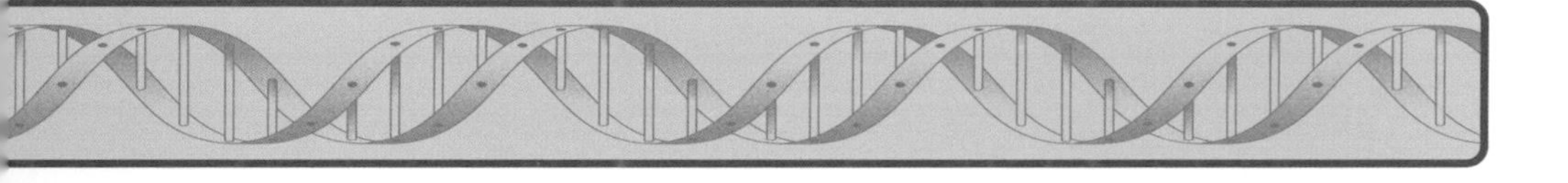

Foreword

The processes that eventually led to life began inside the first generation of stars that resulted after what astrophysicists refer to as the Big Bang. The events associated with the Big Bang mark the beginning of our universe—a time during which such simple elements as hydrogen and helium were turned by gravitational pressure and heat into carbon, oxygen, nitrogen, magnesium, chlorine, calcium, sodium, sulfur, phosphorous, iron, and other elements that would make the formation of the second generation of stars and their planets possible. The most familiar of these planets, our own planet Earth, would give rise to life as we know it—from cells and giant squids to our own human race.

At every step in the processes that led to the rich variety of life on Earth, the thing that was forming in any particular environment was capable of transforming, and did, transform that environment. Especially effective at transformation was that class of things that we now refer to as replicators. We know of two examples of replicators: genes and memes. (The latter rhymes with "creams.")

Genes and memes exist in cells, tissues, and organs. But memes mostly are in brains, and human religions and civilizations. Near the bottom of the hierarchy it's all genes and near the top it's more memes. Genes appeared independently of cells, and are responsible for most of what we call biological life, which can be thought of as a soft and comfortable vehicle made mostly of cells, and created and maintained by the genes, for their efficient replication and evolution. Amazingly, the existence of replicators is all it takes to explain life on Earth; no grand creation, no intelligent design, no constant maintenance; at first just genetic replicators and natural selection, and as far as we know, just one more thing, which appeared after there were human brains big enough to support them: memes.

Genes you've already heard of; but memes may be a completely new term to you. Memes follow the same rules as genes and their natural selection and evolution account for everything that the natural selection of genes doesn't. For instance, our brains are almost too large for our upright stance and therefore must somehow answer to a calling other than the mere replication of our genes, which were doing okay without the extra pint of white matter we gained in the last 50,000 years. The striking increase in brain size means something powerful is strongly benefiting from our increased brain capacity. The best explanation for this, according to Richard Dawkins, in his best-selling and robustly influential book, *The Selfish Gene*, is that our brains are particularly well adapted for imitation, and therefore for the replication of memes. Memes are things like words, ideas, songs, religious or political viewpoints, and nursery rhymes. Like genes, they exist for themselves—that is, they are not here to promote us or anything else, and their continued existence does not necessarily depend on their usefulness to anything: only to their fecundity, their ability to copy themselves in a very precise, but inexact way, and their relative stability over time. It is these features of **genes** and of **memes** that allow them to take part in natural selection, as described by Darwin in 1859, in spite of the fact that he was unaware of the nature of the two replicators. After 150 years, we have started to understand the details. Looking back on it from only a century and a half, Darwin's conception was probably the most brilliant that mankind has chanced upon in our relatively short time here on Earth. What else could possibly explain dandelions?

Dawkins realized that the genes were evolving here, not us. We are just the vessel, and Dawkins realized the significance of replicators in general. After that, the field opened up rather widely, and must include Stan Cohen and Herbert Boyer, whose notion, compounded in 1973 in a late-night deli in Oahu, of artificially replicating specific genes underlies most of the subject matter in this rather important series of books.

SO WHAT'S SO IMPORTANT ABOUT GENES AND MEMES?

I'm sure that most of you might want to know a little bit about the stuff from which you are made. Reading the books in this series will teach you about the exciting field of biotechnology and, perhaps, most importantly, will help you understand what you are (now pay attention, the following clause sounds trivial but it isn't), *and* give you something very catchy to talk about with others, who will likely pass along the information to others, and so on. What you say to them may outlive you. Reading the books in this series will expose you to some highly contagious memes (recall that memes are words, ideas, etc.) about genes. And you will likely spread these memes, sometimes without even being aware that you are doing so.

THE BIRTH OF BIOTECHNOLOGY—OAHU 1973

So what happened in Oahu in 1973? Taking the long view, nothing really happened. But we rarely take the long view, so let's take the view from the 70s.

At the time, it was widely held that genes belonged to a particular organism from whose progenitors the gene had been passed to an organism and that the organism in question would pass the gene to its offspring, and that's the only way that genes got around. It made sense. Genes were known by then to carry the instructions for building new organisms out of the germinal parts of old organisms, including constructing a wide array of devices for collecting the necessary raw materials needed for the process from the environment; genes were the hereditary mechanism whereby like begat like, and you looked like your parents because of similar genes, rather than looking like your neighbors. The "horizontal transfer" of genes from one species to another was not widely contemplated as being possible or desirable, in spite of the fact that such transfer was already evident in the animal and plant worlds—think of mules and nectarines. And "undesirable" is putting it rather mildly. A lot of people thought it was a horrible idea. I was a research scientist in the recombinant laboratories of Cetus Corporation in 1980, during which time Cetus management prudently did not advertise the location of the lab for fear that the good people of Berkeley, California (a town known for its extreme tolerance of most things) might take offense and torch our little converted warehouse of a lab. Why this problem in regard to hybrid life forms? Maybe it had something to do with the fact that mules were sterile and nectarines were fruit.

Apples have been cultivated in China for at least 4,000 years. The genetic divergence from the parental strains has all been accomplished by intentional cultivation, including selection of certain individuals for properties that appealed to our farming ancestors; and farmers did so without much fanfare. The Chinese farmers were not aware that genes were being altered permanently and that was the reason that the scions from favored apple trees, when grafted onto a good set of roots, bred true. But they understood the result. Better apple genes have thus been continually selected by this process, although the process throughout most of history was not monitored at the genetic level. The farmers didn't have any scary words to describe what they were doing, and so nobody complained. Mules and nectarines and Granny Smith apples were tolerated without anyone giving a hoot.

Not so when some educated biologists took a stab at the same thing and felt the need to talk about it in unfamiliar terms to each other, but not the least to the press and the businessmen who were thinking about buying in. There was, perhaps, a bit too much hyperbole in the air. Whatever it was, nobody was afraid of apples, but when scientists announced that they could move a human gene into a bacterium, and the bacterium would go on living and copying the gene, all hell broke loose in the world of biology and the sleepy little discipline of bioethics became a respectable profession. Out of the settling dust came the biotechnology industry, with recombinant insulin, human growth hormone, erythropoietin, and tissue plasminogen activator, to name a few.

CETUS IN 1980

The genie was out of the bottle. Genes from humans had been put into terrified bacteria and the latter had survived. No remarkable new bacteremias—that is, diseases characterized by unwanted bacteria growing in your blood—had emerged, and the initial hesitancy to do recombinant DNA work calmed down. Cetus built a P-3, which was something like an indoor submarine, with labs inside of it. The P-3 was a royal pain to get in and out of; but it had windows through which potential investors could breeze by and be impressed by the bio-suited scientists and so, just for the investment it encouraged, it was worth it. Famous people like Paul Berg at Stanford had warned the biotech community that we were playing with fire. It stimulated investment. When nobody died bleeding from the eyeballs, we started thinking maybe it wasn't all that scary. But there was something in the air. Even the janitors pushing their brooms through the labs at night and occasional scientists working until dawn, felt that something new and promising was stirring.

My lab made oligonucleotides, which are little, short, single-stranded pieces of DNA, constructed from the monomers A, C, T, and G that we bought in kilogram quantities from the Japanese, who made them from harvested salmon sperm (don't ask me how). We broke these DNA pieces down into little nucleoside constituents, which we chemically rebuilt into 15- to 30-base long sequences that the biologists at Cetus could use to find the big pieces, the genes, that coded for things like interferons, interleukins, and human proteins.

We were also talking about turning sawdust into petroleum products. The price of petroleum in the world was over $35 a barrel, if my memory serves me at all, which was high for the decade. A prominent oil company became intrigued with the sawdust to petroleum idea and gave us somewhere between $30 or $40 million to get us started on our long-shot idea.

The oil company funding enabled us to buy some very expensive, sensitive instruments, like a mass spectrometer mounted on the backside of a gas chromatograph, now called GCMS. It was possible under very special conditions, using GCMS, to prove that it could be done—glucose could be converted biologically into long chain hydrocarbons. And that's what gasoline was, and sawdust was mainly cellulose, which was a polymer of glucose, so there you have it. Wood chips into gasoline by next year. There were a few details that have never been worked out, and now it has been a quarter of a very interesting century in which the incentive, the price of oil, is still very painful.

My older brother Brent had gone to Georgia Tech as had I. He finished in chemical engineering and I in chemistry. Brent worked for a chemical company that took nitrogen out of the air and methane out of a pipe and converted them into just about anything from fertilizer to the monomers needed to make things like nylon and polyethylene. Brent and I both knew about chemical plants, with their miles of pipes and reactors and about a century of good technical improvements, and that the quantities of petroleum products necessary to slake the global appetite for dark, greasy things would not fit easily into indoor submarines. We had our doubts about the cellulose to oil program, but proteins were a different thing altogether. Convincing bacteria, then later yeasts and insect cells in culture, to make human proteins by inserting the proper genes not only seemed reasonable to us but it was reasonable.

WE DID IT!

I remember the Saturday morning when David Mark first found an *E. coli* clone that was expressing the DNA for human beta-interferon using a P32 labeled 15-base long oligonucleotide probe that my lab had made. Sometimes science is really fun. I also remember the Friday night driving up to my cabin in Mendocino County when I suddenly realized you could make an unlimited amount of any DNA sequence you had, even if what you had was just a tiny part of a complex mixture of many DNAs, by using two oligos and a polymerase. I called it Polymerase Chain Reaction. The name stuck, but was shortened to PCR.

We were down in a really bad part of town, Emeryville being the industrial side of Berkeley; but we were young and brave, and sometimes it was like an extended camping trip. There were train tracks behind our converted warehouse. You could walk down them during the daytime to an Indian restaurant for lunch, or if you could manage to not be run over by a train late at night while a gel was running or an X-ray plate was exposing, you could creep over across the tracks to the adjacent steel mill and watch white hot steel pouring out of great caldrons. In the evenings, you could go up on the roof and have a beer with the president of the company.

Like the Berkeley of the late Sixties which had preceded it, it was a time that would never happen again.

Today, nobody would be particularly concerned about the repercussions of transferring a gene out of bacteria, say a gene out of *Bacillus thuringiensis* inserted into a commercial strain of corn, for instance. Genes now have found a new way to be moved around, and although the concept is not revolutionary, the rate at which it is happening is much faster than our own genes can react to. The driver, which is the case for all social behaviors in humans today, is the meme. Memes can appear, replicate, and direct our actions as fast as thought. It isn't surprising, but it does come as a shock to many people when they are confronted with the undisputed fact that the evolving elements in what we have referred to as biological evolution, which moves us from *Homo habilus* to *Homo sapiens*, are genes; not organisms, packs, species, or kinship groups. The things that evolve are genes, selfishly. What comes as an even more shocking surprise, and which in fact is even less a part of the awareness of most of us, is that our behavior is directed by a new replicator in the world, the meme.

YOU MAY WANT TO SKIP THIS PART

(Unless You're Up for Some Challenging Reading)

Let's digress a little, because this is a lot of new stuff for some people and may take a few hours to soak in. For starters, what exactly is a gene? . . . atgaagtgtgccgtgaaagctgctacgctcgacgctc-gatcacctggaaaaccctggtag . . . could be the symbol for a gene, a rather short gene for our editorial convenience here (most of them have thousands of letters). This rather short gene would code for the peptide met-lys-cys-ala-val-lys-gly-gly-thr-leu-asp-ala-arg-ser-pro-gly-lys-pro-trp, meaning that in a cell, it would direct the synthesis of that string of amino acids, (which may or may not do something very important).

Getting back to the gene, it may share the organism as an environment favoring its replication with a whole gang of other replicators (genes), and they may cooperate in providing a comfy little protected enclave in which all of the genes develop a means to replicate and cast their sequences into the future all using the same mechanism. That last fact is important as it separates a cellular gene from a viral gene, but I won't belabor it here. Reviewing just a little of what I've infected you with, that sequence of AGCT-type letters above would be a replicator, a gene, if it did the following:

(1) exhibited a certain level of fecundity—in other words, it could replicate itself faster than something almost like it that couldn't keep up;

(2) its replication was almost error-free, meaning that one generation of it would be the same as the next generation with perhaps a minor random change that would be passed on to what now would be a branch of its gene family, just often enough to provide some variation on which natural selection could act; and

(3) it would have to be stable enough relative to the generation time of the organism in which it found itself, to leave, usually unchanged, with its companion genes when the organism reproduced.

If the gene goes through the sieve of natural selection successfully, it has to have some specific identity that will be preserved long enough so that any advantage it confers to the rate of its own replication will, at least for some number of generations, be associated with its special identity. In the case of an organic replicator, this specialness will normally be conferred by the linear sequence of letters, which describe according to the genetic code, a linear sequence of amino acids in a peptide. The process is self-catalytic and almost irreversible, so once a sequence exhibits some advantage in either (1), (2), or (3), all other things being nearly equal, it is selected. Its less fortunate brethren are relatively unselected and the new kid on the block takes over the whole neighborhood. See how that works?

This should not be shocking to you, because it is a tautology, meaning it implies nothing new. Some people, however, are accustomed to the notion that genes and individual organisms serve the greater good of something they call a species, because in the species resides an inviolate, private gene pool, which is forever a part of that species. This concept, whether you like it or not, is about as meaningful—and now I guess I will date myself—as the notion that Roger Waters is forever and always going to be playing with Pink Floyd. It isn't so. Waters can play by himself or more likely with another group. So can genes. And don't forget that not only genes, but also an entirely different kind of replicator is currently using our bodies as a base of operations. Genes have a reaction time that is slow relative to the lifetime of an individual. It takes a long time for genes to respond to a new environment. Memes can undergo variation and selection at the speed of thought.

Let's leave the subject of memes for awhile. They are an immense part of every human now, but biotechnology as practiced in the world and described in this book does not pay them much mind. Biotechnologists are of the impression that their world is of genes, and that's alright. A whole lot happened on Earth before anybody even expected that the place was spinning and moving through space, so memes can wait. I thought I ought to warn you.

It is worth noting that new gene *sequences* arise from preexisting gene sequences but gene *molecules* are not made out of old genes. Gene molecules are made out of small parts that may have been in genes before, but the atoms making up the nucleotides that are strung together and constitute today's incarnation of a gene, may have two weeks ago been floating around in a swamp as urea or flying out of a volcano as hot lava. A gene *sequence* (notice that molecule is not equal to sequence) that makes itself very useful may last millions of years with hardly a single change. You may find precisely the same gene sequence in a lot of very different species with few significant changes because that sequence codes for some protein like cytochrome C that holds an iron atom in a particularly useful way, and everybody finds that they need it. It's a more classic design than a Jaguar XK and it just keeps on being useful through all kinds of climatic eras and in lots of different species. The sequence is almost eternal. On the very different other hand, the specific molecular incarnations of a gene sequence, like the DNA molecule that encodes the cytochrome C sequence in an individual cell of the yeast strain that is used, for example, to make my favorite bread, Oroweat Health Nut, is ephemeral. The actual molecules strung together so accurately by the DNA polymerase to make the cytochrome C are quickly unstrung in my small intestine as soon as I have my morning toast. I just need the carbon, nitrogen, and phosphorous. I don't eat it for the sequence. All DNA sequences taste the same, a little salty if you separate them from the bread.

That's what happens to most chemical DNA molecules. Somebody eats them and they are broken down into general purpose biological building blocks, and find their way into a new and different molecule. Or, as is often the case in a big organism like Arnold Schwarzenegger, body cells kill themselves while Governor Schwarzenegger is still intact because of constantly undergoing perfectly normal tissue restructuring. Old apartments come down, new condos go up, and beautiful, long, perfectly replicated DNA sequences are taken apart brick by brick. It's dangerous stuff to leave around on a construction site. New ones can be made. The energy just keeps coming: the sun, the hamburgers, the energy bars.

But the master sequences of replicators are not destroyed. Few germ cells in a woman's ova and an embarrassingly large number of germ cells in a male's sperm are very carefully left more or less unaltered, and I say more or less, because one of the most important processes affecting our genes, called recombination, does alter the sequences in important ways; but I'm not going to talk about it here, because it's pretty complicated and this is getting to be too long. Now we are ready to go back to the big question. It's a simple answer, but I don't think you are going to get it this year.

If . . . atgaagtgtgccgtgaaagctgctacgctcgacgctcgatcacctggaaaaccct-ggtag . . . is a gene, then what particular format of it is a gene? For this purpose, let's call it a replicator instead of a gene, because all genes are after all replicators. They happen to encode protein sequences under certain conditions, which is one of our main uses for them. As I've mentioned, the one above would code for the protein met-lys-cys-ala-val-lys-gly-gly-thr-leu-asp-ala-arg-ser-pro-gly-lys-pro-trp with the final "tag" being a punctuation mark for the synthesis mechanism to stop. We make other uses of them. There are DNA aptamers, which are single-stranded DNA polymers useful for their three-dimensional structures and ability to specifically cling to particular molecular structures, and then there is CSI, where DNA is used purely for its ability to distinguish between individuals. But *replication* for the genes is their reason for being here. By "reason for being here," I don't mean to imply that they are here because they had some role to fulfill in some overall scheme; I just mean simply that they are here because they replicate—it's as simple or impossible to understand as that. Their normal way of replication is by being in their molecular form as a double stranded helical organic polymer of adenosine, guanosine, thymidine, and cytosine connected with phosphate linkages in a cell. Or they could be in a PCR tube with the right mixture of nucleoside triphosphates, simple inorganic salts, DNA polymerase, and short strands of single-stranded DNA called primers (we're getting technical here, that's why you have to read these books). Looking ahead, DNA polymerase is a molecular machine that hooks the triphosphate form of four molecular pop beads called A, C, T, and G together into long meaningful strings.

Okay, getting back to the question, is the gene, . . . atgaagtgtgccgtgaaagctgctacgctcgacgctcgatcacctggaaaaccctggtag . . . always the organic polymer form of the sequence, which has a definite mass, molecular weight, chemical structure, or is the Arabic letter form of it in your book still a gene, or is the hexadecimal representation of it, or the binary representation of it in your CPU a gene, or is an equivalent series of magnetic domains aligned in a certain way on your hard disk just another form of a gene? It may sound like a dumb question, but it isn't. If you are insistent that a gene is just the organic polymer of A, C, T, and G that can be operated on by DNA polymerase to make replicas in a cell, then you may take a minute to think about the fact that those little triphosphate derivatives of A, C, T, and G may not have been little nucleotides last month when they were instead disembodied nucleotide pieces or even simple atoms. The atoms may have been residing in things called sugars or amino acids in some hapless organism that happened to become food for a bigger organism that contained the machinery that assembled the atoms into nucleotides, and strung them into the sequence of the gene we are talking about. The thing that is the same from generation to generation is the sequence, not the molecule. Does that speak to you? Does it say something like maybe the symbol of the gene is more the gene than the polymer that right now contains it, and the comprehensive symbolic representation of it in any form at all is a replicator? This starts to sound pretty academic, but in any biotechnology lab (and you will read about some of them in this series) making human proteins to sell for drugs, the genes for the proteins take all the above mentioned forms at one time or another depending on what is appropriate, and each of them can be reasonably called a replicator, the gene. Genetic engineering is not just the manipulation of chemicals.

SKIP DOWN TO HERE

These books are not written to be the behind the scenes story of genes and memes any more than a description of an integrated circuit for someone who wants to use it in a device for detecting skin conductivity or radio waves is about quantum mechanics. Quantum mechanics is how we understand what's happening inside of a transistor embedded in an integrated circuit in your iPod or described in the Intel catalogue. By mentioning what's going on inside biotechnology, I hope to spark some interest in you about what's happening on the outside, where biotechnology is, so you can get on about the important business of spreading these memes to your friends. There's nothing really thrilling about growing bacteria that make human hormones, unless your cousin needs a daily injection of recombinant insulin to stay alive, but the whole process that you become involved in when you start manipulating living things for money or life is like nothing I've found on the planet for giving you the willies. And remember what I said earlier: you need something to talk about if you are to fulfill your role as a meme machine, and things that give you the willies make great and easily infectious memes. Lowering myself to the vernacular for the sake of the occasional student who has made it this far, "biotech is far out man." If you find something more interesting, let me know. I'm at kary@karymullis.com, usually.

Dr. Kary B. Mullis
Nobel Prize Winner in Chemistry, 1993
President/Altermune, LLC

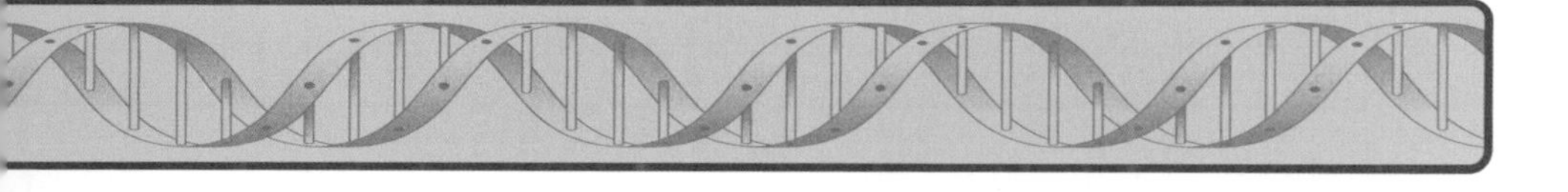

Introduction

Biotechnology, the use of biological organisms and processes to provide useful products in industry and medicine, is as old as cheese making and as modern as creating a plant-based energy cell or the newest treatment for diabetes. Every day, newspaper articles proclaim a new application for biotechnology. Often, the media raises alarms about the potential for new kinds of biotechnology to harm the environment or challenge our ethical values. As a result of conflicting information, sorting through the headlines can be a daunting task. These books are designed to allow you to do just that—by providing the right tools to help you to make better educated judgments.

The new biotechnologies share with the old a focus on helping people lead better, safer, and healthier lives. Older biotechnologies, such as making wine, brewing beer, and even making bread, were based on generations of people perfecting accidental discoveries.

The new biotechnologies are built on the explosion of discoveries made over the last 75 years about how living things work. In particular, how cells use genetic material to direct the production of proteins that compose them, and provide the engines used to produce energy needed to keep them alive. These discoveries have allowed scientists to become genetic engineers, enabling them to move genes from one living organism to another and change the proteins made by the new organism, whether it is a bacterium, plant, mouse, or even a human.

Biotechnologists first engineered bacterial cells, producing new proteins useful in medicine and industry. The type of cell that biotechnologists engineer today may be a simple bacterium or a complicated animal or human cell. The protein product might be a simple string of amino acids or a complicated antibody of four chains, with critical genetic instructions from both mice and humans. Plant biotechnologists engineer plants to resist predatory insects or harmful chemicals to help farmers produce more, with less risk and expense. Plants have also been engineered to make products useful for industry and manufacturing. Animals have been engineered for both research and practical uses.

Research is also underway to develop methods of treating human diseases by changing the genetic information in the cells and tissues in a patient's body. Some of these efforts have been more successful than others and some raise profound ethical concerns. Changing the genetic information of a human may one day prevent the development of disease, but the effort to do so pushes the envelope of both ethics and technology. These and other issues raised by advances in biotechnology demand that we as citizens understand this technology, its promise, and its challenge so that we can provide appropriate limits on what biotechnologists create.

Who are the biotechnologists, the genetic engineers? Generally, they have university or advanced training in biology or chemistry. They may work in a university, a research institute, a company, or the government. Some are laboratory scientists trained in the tools of genetic engineering—the laboratory methods that allow a gene for a particular protein to be isolated from one living creature's DNA and inserted into another's DNA in a way that instructs the new cell to manufacture the protein. Some are computer scientists who assemble databases of the DNA and protein sequences of whole organisms. They may write the computer code that allows other scientists to explore the databases and use the information to gain understanding of evolutionary relationships or make new discoveries. Others work in companies that engineer biological factories to produce medicines or industrial plastics. They may engineer plants to promote faster growth and offer better nutrition. A few have legal training that allows them to draft or review patents that are critical to the business of biotechnology. Some even work in forensic laboratories, processing the DNA fingerprinting you see on TV.

The exact number of working biotechnologists is hard to determine, since the job description doesn't neatly fit into a conventional slot. The U.S. Department of Labor indicates that there are over 75,000 Master's and Ph.D.-level biologists in the U.S., and Bio.org, the Website for the Biotechnology Industry Organization, reports nearly 200,000 biotechnologists are currently employed.

Biotechnology is not just the stuff of the future. The work of modern biotechnology and genetic engineering is in our daily lives, from the food we eat and clothing we wear to some of the medicines we take. The ketchup you put on your fries at lunch today may have been sweetened with corn syrup made from corn that was engineered to resist a deadly insect. The cotton in your T-shirt, even if the shirt were made in China or Bangladesh, probably came from a U.S. cotton plant genetically engineered to resist another insect. If someone in your family is a diabetic, the insulin he or she injects and the glucose test monitor used to determine the amount of insulin to inject rely on biotechnology. If you go to the doctor and she arranges for blood tests, the laboratory uses biotechnology products to run those tests.

This series, BIOTECHNOLOGY IN THE 21ST CENTURY, was developed to allow you to understand the tools and methods of biotechnology and to appreciate the current impact and future applications of biotechnology in agriculture, industry, and your health. This series also provides an exploration of how computers are used to manage the enormous amount of information produced by genetic researchers. The ethical and moral questions raised by the technology, whether they involve changing the genetic information of living things or using cells from human embryos to develop new ways to treat disease, are posed with a foundation in how moral philosophers think about ethical issues. With these tools, you will be better able to understand the headlines about the latest advances in biotechnology and the alarms raised by those concerned with the impact that these applications have on the environment and our society. You may even be inspired to learn more and join the community of scientists who work on finding new and better ways to produce food, products, and medical treatments.

Bernice Zeldin Schacter, Ph.D.
Consulting Editor

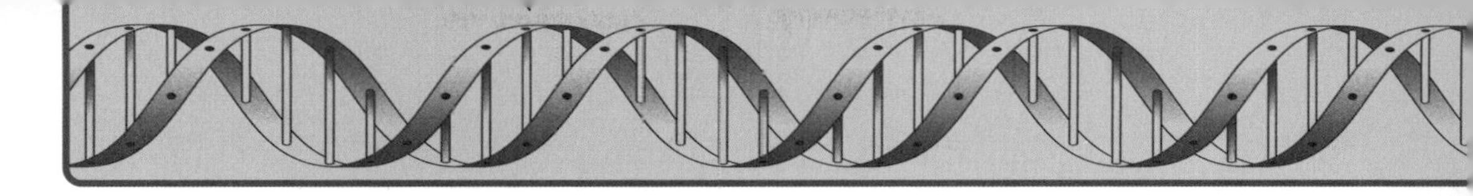

1

DNA, Proteins, and Cells

This book is about something so new that it is revolutionizing biology. **Bioinformatics** refers to the generation and mining of computerized databases of hereditary information. **Genomics** has to do with identifying the entire DNA code of not just one living thing, but comparing the complete DNA codes of many living things. **Proteomics** refers to new, fast, and efficient ways to identify, compare, and study many **proteins** at one time. All of these tools are changing the way that scientists go about exploring the inner workings of cells. These emerging fields of study are also changing the ways that ailments from cancer to schizophrenia are diagnosed and are driving the development of new drugs that target specific defective proteins running amok and causing disease. Similarly, these tools provide scientists with more effective ways to engineer plants that can kill their predators, **microbes** that can digest pollution, and fish that will glow in your tank. Indubitably, databases of

genome sequences and protein structure, the computer tools we have to mine them, and the **molecular genetics**, chemical engineering, and clever robotics that allow us to explore molecular interactions on a genomic scale are dramatically altering the way that we think about our relationship to all living things.

STUART SCHREIBER'S EPIPHANY

Watching the male cockroaches stand up on their hind legs and beat their wings in a frenzy of anticipation, Professor Stuart Schreiber was transfixed. This young faculty member at Yale University had just released a pure elixir of concentrated molecules into the air, copied from a love-potion, a molecule honed by generations of female cockroaches to drive their mates wild. But the real epiphany came when he applied the aphrodisiac after first attaching the roach's antenna to a voltmeter. The voltmeter could indicate a change in the electrical charge inside the antenna. The voltmeter would tell him if the chemistry of the cell was changing in response to the love-potion he put on the antenna. When he saw the needle move, according to Stephen Dickman in *Exploring the Biomedical Revolution*, Schreiber said, "I thought to myself, that's just chemistry. There must be a receptor in the cell membrane, communicating with something inside the cell."[1]

Schreiber's imagination had made an amazing leap. Trained to think like the chemist he was and the high school shop enthusiast he had been not too long ago, he assumed that his love-potion molecule must be sliding into a cranny, just as a key slips into a lock, in some molecule on the roach's antenna. How could this **binding** event elicit the roach's inevitable and frantic dance? Stuart assumed that the roach receptor molecule, a protein whose shape had evidently been altered by the attachment of the aphrodisiac, must affect the activity of other molecules, possibly setting off a relay of such events. This unknown cascade of "contact, engage, and power up" must have changed the activity of the roach's muscles *inside* the roach's body, causing him to move! (Figure 1.1)

Stuart Schreiber's main interests in high school had been shop, cars, girls, and having a good time. It was only through a similar moment of insight, sitting in on his first chemistry class at the University of Virginia, that Schreiber was enticed by the simple beauty and regularity of **electron clouds**, the probable sites for finding electrons around an atom, drawn on the blackboard by his professor. Perhaps Schreiber transferred his fascination with cars and their interconnected parts to understanding how the atomic composition and shapes of small molecules determined their inclination to bind to other molecules. Although he was on the verge of dropping out of school, he decided *not* to drop out and instead pursued his new interest in chemistry with such enthusiasm that at age 26 he became the youngest full professor in chemistry at Yale—*ever*. But his later success in manufacturing the molecules of roach aphrodisiac simulated another career change—using chemistry as a tool to learn about biology.

BAGS OF GOOP

You are probably aware that all living things (often called **organisms** by scientists) are made up of one to many tiny cells. Each cell is bounded by at least one greasy bag, called the **plasma membrane.** Inside this membrane is a watery gelatin-like substance interspersed with (in every organism more complicated than a bacterium) smaller bags, each containing their own watery goop. Amazingly, many of these smaller bags of goop, or **organelles**, are quite similar in organisms as small as yeast and as complex as a spruce tree or a human being. Each organelle has a special function. The large molecules that do most of the work inside the cell and within these organelles are **proteins**, which function with the precision of tiny machines. Many of these proteins activate other molecules when a particular molecule binds to them, just as Schreiber

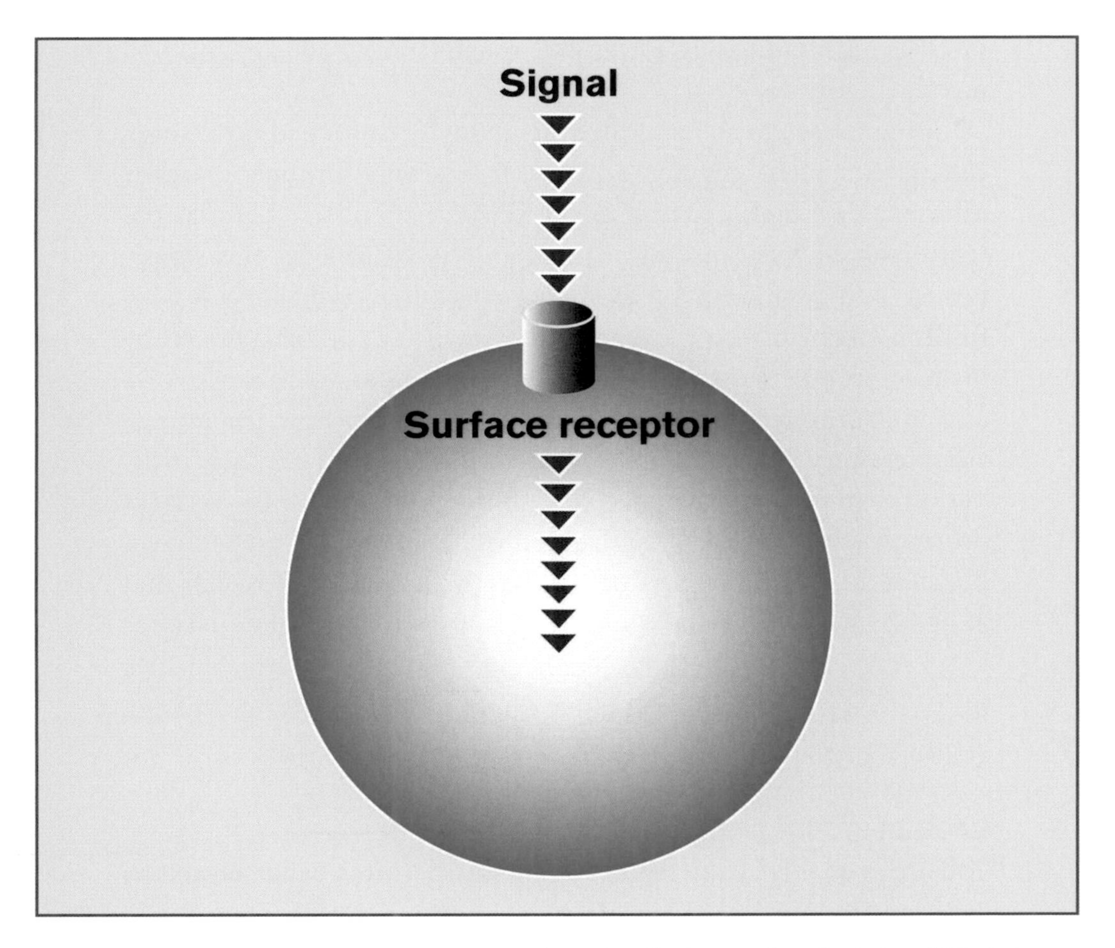

Figure 1.1 Cells are able to respond to outside stimuli when a chemical signal binds to a surface receptor. That binding event may start a cascade of events that ultimately changes what happens inside the cell. In some cases, the binding event may open a channel that starts a signal. Other times, the binding event may activate a portion of a transmembrane protein so that it chemically modifies another protein within the cell, initiating a cascade of events.

imagined, or they act as **enzymes**, proteins that help other molecules stay together long enough to make a connection or break apart. Some proteins require chemical energy packets called ATP (Adenosine triphosphate) to do their specific jobs.

It is important to note that the proteins made by an organism determine all of the characteristics that "nature" provides for that particular living thing. (Interactions with the environment, such as eating good food, seeing your friends, and working hard in school provide the "nurture" part of the equation that can also change the chemistry of your body.) The enzymes allow other molecules, including **proteins**, **fats**, and **carbohydrates** to undergo chemical reactions, such as being put together or taken apart inside living things. In addition, some proteins make up structural elements such as surface cell receptors, like the one that recognizes the AIDS virus and allows it to enter some of your blood cells called T cells. One percent of Caucasians do not have this receptor and are therefore immune to the HIV virus that causes AIDS.[2] Other proteins or small molecules interact with certain receptors to pass a signal from the surface of the cell to its interior and onto other cells by a relay of events similar to those that changed the aphrodisiac signal into roach muscle movement and wing flapping. Other proteins bind DNA, the molecules of heredity, and determine which codes are going to be used to make proteins—at what time and in which type of cell.

Because each protein has an important job to do, it is crucial that proteins be made to precise specifications, just like the precision parts of an expensive sports car. In fact, the blueprints for some proteins have been so good, they have been preserved through millions and even billions of years of evolution. This is true for proteins that carry out basic processes in the cell, like the protein machine called the ATP synthase that makes the energy molecule ATP (Figure 1.2). The ATP synthase, like many tiny protein machines, is made up of several long protein chains folded around each other. These protein chains are practically identical in chimpanzees and humans. One of the ATP synthase protein chains from a yeast (the fungus that we use in making beer and bread) is over 70% identical to that in the human ATP synthase. Some proteins are so similar from one living organism to the next that

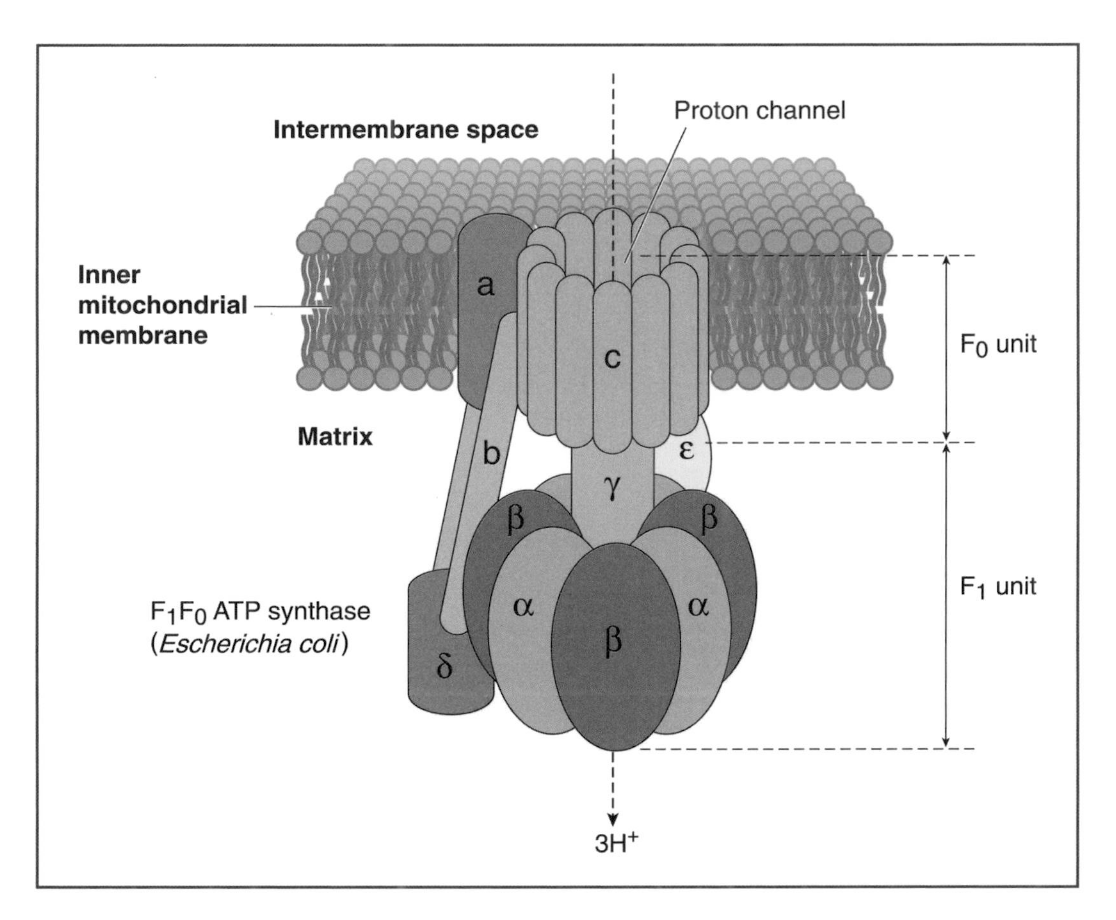

Figure 1.2 The structure of the ATP synthase is illustrated here. This universal machine that combines ADP with inorganic phosphate to make ATP is made up of multiple protein subunits. As hydrogen ions pass from one side of the inner mitochondrial membrane to the other, they cause the F_0 portion of this protein machine to turn. This makes the central shaft turn, deforming the alpha-beta bulbs of the F_1 portion that squeeze ADP and phosphate together.

they can literally substitute for that protein machine in the cell. For example, a **proton pump** in a tiny plant called *Arabiodopsis thaliana* pushes hydrogen ions out of plant cells so that nutrients can be escorted in with the hydrogen ions clamoring to get back in. The proton pump in the yeast plasma membrane functions in the same way to generate a hydrogen gradient, a difference in the concentrations of hydrogen ions on either side of the membrane, that will power the import of nutrients. When the yeast gene for the proton pump is substituted with the plant variety, the plant pump, with a slight modification, can work in yeast and keep the cells alive. (Cells without the yeast proton pump normally die.)

THE LIFESAVERS™ MODEL OF DNA

The blueprints for all of our proteins are carried in each of our cells. A very important organelle or bag of goop in each cell is the **nucleus.** This small structure contains molecules of **DNA**, long string-like molecules that contain a simple, but rich, informational code for directing the production of all the specific proteins in the cell. Each protein is made up of either a single chain or several chains of **subunits** called **amino acids**. In contrast, DNA is made up of a long string of subunits called **bases**. It is important not to confuse DNA and protein. They are two very different molecules.

Our cells are able to read the DNA code in the nucleus and produce a molecule called messenger RNA (mRNA) that carries that code in a special way. The mRNA molecule travels outside of the nucleus to direct the stringing together of amino acids into particular chains that are either themselves proteins or form a portion of a protein. Therefore the DNA code is **transcribed** into an mRNA code which is then **translated** into protein. As stated above, proteins often act as enzymes that allow particular reactions to take place; for example, to make a brown or blue pigment in our eyes. Therefore, DNA codes for producing proteins that, along with our experiences, determine our characteristics. Particular sections of our DNA molecules provide the code for each string of amino acids. These sections are called **genes**. The original copies of these DNA molecules came in the chromosomes in the egg and sperm from our mother and father (Figure 1.3).

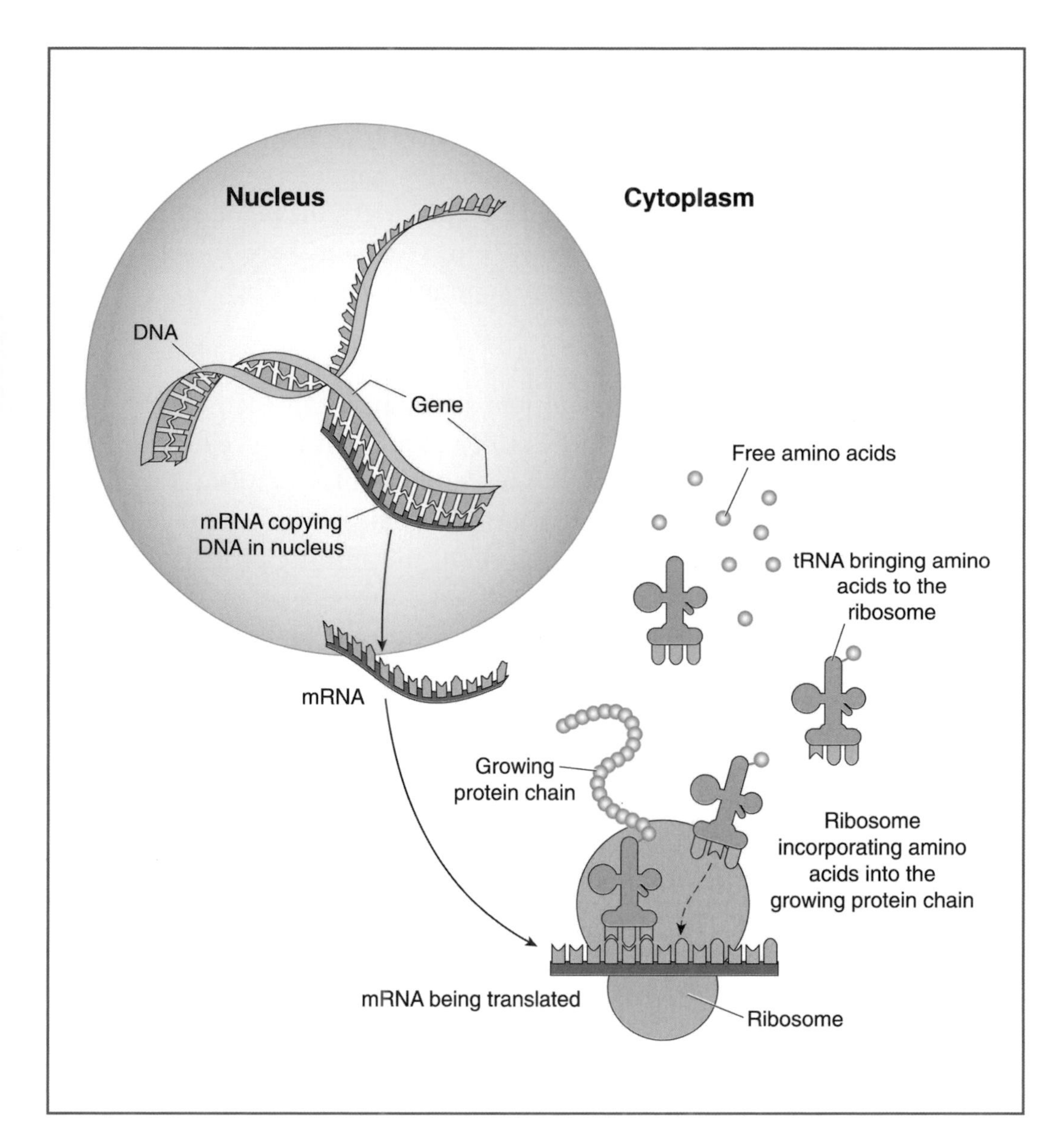

Figure 1.3 The process by which DNA codes for a protein is illustrated here. One side of a portion of the DNA code in the nucleus is copied into mRNA. The mRNA moves out of the nucleus to a ribosome in the cytoplasm. There, tRNAs bring individual amino acids one at a time to be added to a growing chain of amino acids that is a protein.

Chromosomes are the dark structures inside each cell's nucleus. Each chromosome contains both protein and one long, thin molecule of DNA. In the late 1800s and early 1900s, scientists thought that chromosomes might be involved in heredity. They followed the structures of duplicated chromosomes by looking at their movements using microscopes. If new cells were going to have the same genetic information, it was important that the parental cell made an extra set of each of its chromosomes before dividing. The chromosomes would then line up across the middle of the cell. Miraculously they would split into two groups as they moved in concert to each end of the cell. When the cell finally finished splitting in two, each new **daughter cell** had the original number of chromosomes. In this way, each cell within a living thing has identical genetic information, except for cells that are eggs or sperm. (This is why DNA left at the scene of a crime, whether it is in hair or blood, for example, can be used to identify the perpetrator.) It is necessary for eggs and sperm to carry half of the total number of chromosomes so that when they combine to make a new individual, that living thing will have the correct number of chromosomes. Although chromosomal DNA is packed into these dark rods with the help of proteins, experiments showed that DNA, not the proteins, are the blueprints for making more proteins.

It is surprising that DNA carries the blueprints for making proteins because DNA is a very simple molecule. DNA molecules are made up of only four types of bases that we can picture as four flavors of LifeSavers—cherry, orange, lemon, and lime. In fact, one can picture the single DNA molecule in one long human chromosome as a *double row* of LifeSavers, each row on average 150 million candies long (see the box on pages 10–11). The two black lines represent the alternating sugar and **phosphate backbones** on the two sides of the DNA molecule. Amazingly, your DNA is almost identical to that of your neighbor across the street and similarly almost identical to that of a student in Botswana or a grandmother in the Ukraine. In other words, the arrangement of the candy flavors is identical, *except* for on average one candy in every thousand counting down one row of LifeSavers. For example, at that

The LifeSavers Model of DNA, Transcription, and Translation

Figure 1.4 shows a part of the double-stranded DNA represented as two stacks of LifeSavers candies. Each LifeSavers flavor corresponds to one of the bases A, T, G, or C. Notice that cherry and orange always pair (G-C) and that lemon and lime always pair (T-A). The dark lines on the outside of the stacks of bases refer to a chemical **backbone** which was described in detail by Watson and Crick. The important feature is that there is a directionality or **polarity** to the orientation of the backbone *and* the bases, just like two pencils that have been aligned opposite each other. The eraser is on the **5'** (pronounced *five prime*) end and the point is on the **3'** (*three prime*) end. It is important to note that the backbones of the **base pairs** always point in opposite directions. (This information will be important later.)

Refer back to Figure 1.3, which shows how the sequence of part of an mRNA strand is determined. The DNA has split open at this point and the bases have come in to line up opposite only one side of the DNA. Note that mRNA uses U instead of T. Therefore, when the mRNA aligns with half of the DNA ladder to momentarily create another ladder, it uses C-G and A-U pairings to create the rungs.

After being created, the mRNA leaves the nucleus and is clamped into place in organelles called the ribosomes. Here individual amino acids are brought in by their special escort molecules called tRNA or transfer RNA that have a 3-base anticodon of nucleotides on their backsides. The anticodons match up with the 3-base codons on the mRNA to bring the particular amino acids into place. There is always only one particular type of amino acid carried by each type of escort.

Next, the protein chain begins to fold. The precise and unchanging folding pattern for each protein (which gives it its unique shape) is determined by the precise sequence of amino acids and also by other proteins called **chaperones** that help the protein to fold correctly. Soon this protein machine will be sent to its proper place in the cell, ready to do its job.

position along one row, your neighbor might have cherry and you might have lime. The student in Botswana and the grandmother in the Ukraine might also have a lime at that position, but their DNA

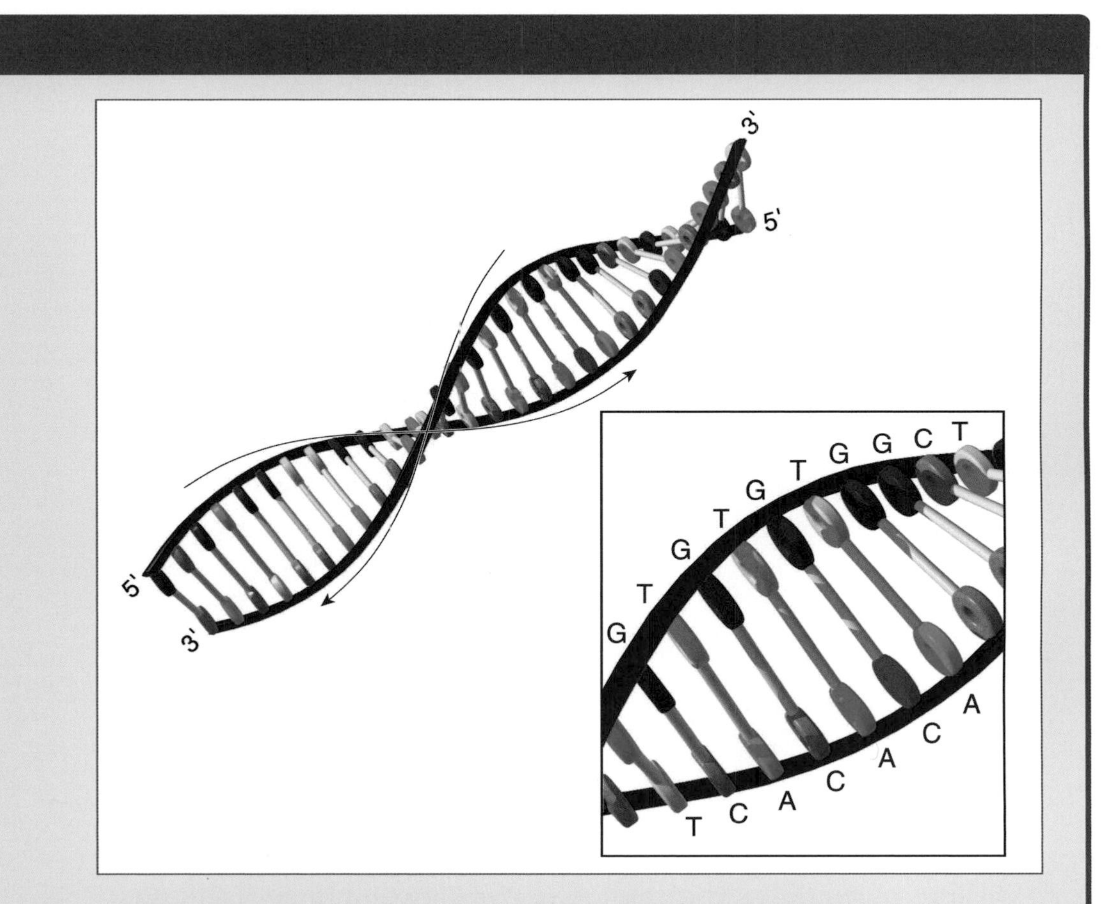

Figure 1.4 One may make a model of DNA by creating two strings of LifeSavers candies. The LifeSavers flavors correspond to the bases A, T, G, and C. Notice that cherry and orange always pair (G-C) and that lemon and lime always pair (T-A) to create the rungs of this spiraled (helical) DNA ladder. The dark lines on the outside of the stacks of bases refer to a chemical backbone made up of alternating sugar and phosphate which was described in detail by Watson and Crick.

might be different from yours at some other position, but at only one out of every thousand candies down each side. If you have an identical twin, your DNA is identical. (Why do you think identical twins look almost exactly alike?)

HOW DOES DNA DUPLICATE ITSELF?

Figuring out which part of the cell contains the code and how that code works provides one of the most compelling stories of science in the 20th century. Experiments between the 1920s and 1950s demonstrated that the cell's coding material was its DNA, not its protein as earlier thought. In 1953, a brash, young Chicagoan named James Watson and his British colleague, Francis Crick, inspired by the work of one of their competitors, Rosalind Franklin, made that code almost self-evident by unveiling the 3-D structure of DNA. In their understated, essentially one-page report in the British journal *Nature*, they wrote, "We wish to suggest a structure for the salt of **deoxyribose nucleic acid** (**D.N.A.**). This structure has novel features which are of considerable biological interest."[3]

Of considerable interest indeed! What they unveiled (in a bit more detail) was a double row of LifeSavers twisted along its length into a spiral or **helix**. Cleverly, they had shown that cherry was always matched with orange and that lemon was always matched with lime, to make the rungs of the willowy spiraled ladder—with absolutely no exceptions allowed. Herein lay the key to the ladder's ability to faithfully duplicate itself. By splitting down the middle into two half ladders and affixing new cherry, orange, lime, and lemon LifeSavers to the exposed sides, two *identical* double strands of LifeSavers could be made. Thus by unzipping the DNA ladder and filling in the missing parts, the DNA was able to faithfully duplicate or **replicate** itself. In this way, the DNA from a single fertilized egg replicates itself before it divides. That first cell and its progeny divide and subdivide so that eventually the new human being, poodle, or rose has identical DNA (and identical chromosomes) in all of its many cells from its big toe (if it has one) to its petals (if it has those). As mentioned before, it is important to remember that eggs and sperm have only half the number of chromosomes, otherwise the number of chromosomes would double with each generation!

Stop and Consider

Why was Watson and Crick's discovery of the molecular structure of DNA so important? How has it helped us to understand what genes are and how they code for proteins?

CONNECTIONS

This chapter has demonstrated the importance of precision protein machines in the functioning of living things. A protein receptor allowed the cockroach to react to the love-potion that Stuart Schreiber put on its antenna. It reacted because of a series of reactions inside the roach mediated by proteins. Other proteins act as receptors that allow the AIDS virus to enter our cells, but a small number of people are resistant because they do not have the DNA blueprint for the viral receptor. Many proteins are enzymes that monotonously perform the same task over and over, but a proper protein structure is crucial for their activity. Enzymes allow for chemical reactions to occur in living things.

The blueprints for all of these proteins are found in the chromosomes of cells. Although the chromosomes are made of both DNA and protein, it is the DNA that carries the hereditary information in the series of four types of bases. Because of the invariant base pairing rules, DNA is able to split down the middle and make an exact copy of itself before the cell divides. Therefore each cell within an individual has the same DNA.

FOR MORE INFORMATION

For more information about the concepts discussed in this chapter, explore the following Websites:

Go to: *www.biointeractive.org* to learn more about Stuart Schreiber's reasons for becoming a scientist and his new ideas about how to use small molecules to learn about biology. Click on the "DNAinteractive" icon. View or order your free copy of the 2002 Howard Hughes Medical Institute Holiday Lecture "Scanning Life's Matrix: Genes, Proteins, and Small Molecules." Stuart Schreiber, Ph.D., of Harvard, and Eric Lander, Ph.D., who is the director of the Whitehead Institute/MIT Center for Genome Research, presented this lecture for high school students during the Christmas holidays in 2002.

Go to: *www.biointeractive.org* to view a fantastic animation in real time of DNA replication. Click the "DNAinteractive" icon. Then click on the tab "Code" and then on the tab "Copying the Code." At each stage you will need to wait to let the movies load. Then click on the tab "Putting it Together" and then on the "Replication" icon.

2

Genes Code for Proteins

Dr. Eric Gouaux, a National Science Foundation young investigator, eloquently described what it would be like to "don the guise of a potassium ion and embark on a trip through the [potassium ion] channel."[1] **Ion channels** are proteins that line holes in the plasma membrane. They can open on demand to let ions in and out of the cell. They allow nerve impulses to travel, cause your heart to beat, and allow your muscles to contract. In many cells, channels and another kind of protein called a pump together maintain a relatively constant negative charge within your cells. This net negative charge, or membrane potential, affects the entry and exit of a variety of materials.

As shown in Figure 2.1, this potassium ion channel looks like an inverted tepee sitting in the cell membrane.[2] Its broad end opens to the outside of the cell and its narrow end is a door that contacts the cell's innards or **cytoplasm**. Once Dr. Gouaux entered

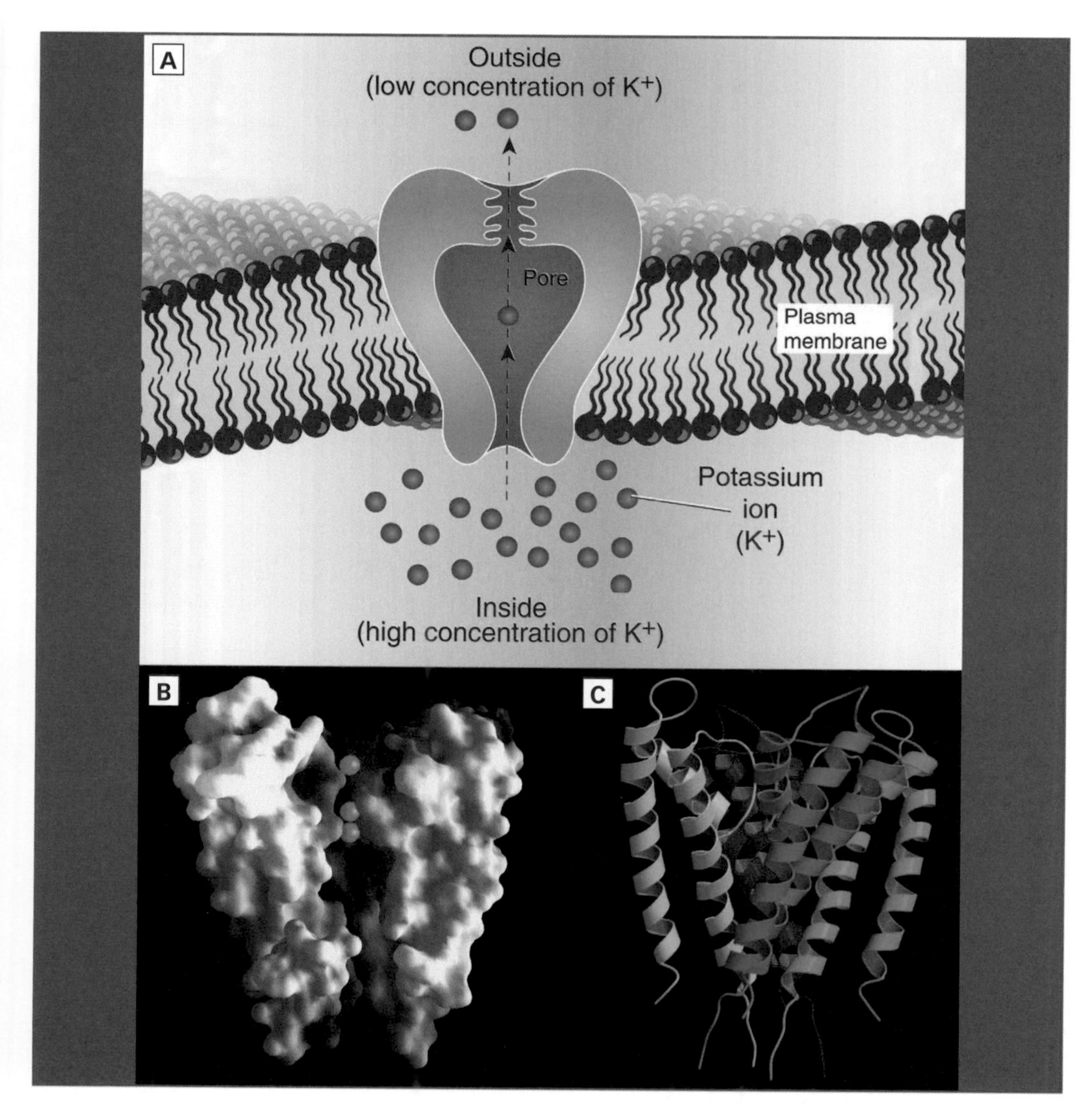

Figure 2.1 A) A model of the potassium channel, which sits in the cell membrane and allows only potassium to pass through it when open. Notice the large central chamber and narrowed area or selectivity filter at the top. B) This is a cutaway view of the actual positions of atoms within the KcsA potassium channel. Three possible positions for potassium ions are shown within the selectivity filter. C) This stylized cartoon view of KcsA shows the contributions of the backbones of the four subunits and the three alpha helices (spiraled areas) within each subunit.

the channel (disguised as K^+, metaphorically), he would feel the tug of the four short **pore helices**, the four short cylinders of protein aimed toward the middle of the channel (Figure 2.1). The negative ends of each cylinder, or helix, would be aimed so as to tug at his single positive charge. After being pulled into the center of the water-soaked vestibule, he would not slow as he entered the narrow path in front of him. A gauntlet of close-fitting **carbonyl oxygen** atoms lining this last portion of the channel would reach out to exchange their arms for most of the water molecules surrounding him and hustle him through. He would move so fast that 10,000,000 to 100,000,000 potassium ions just like him could cross in a second. What a slick way to cross a greasy membrane that would normally be an impenetrable barrier to a positively **charged** ion!

The **potassium ion** channel is made up of four interlocking and identical protein subunits. (Each protein subunit is just one long chain of amino acids folded around itself in a particular way.) Ion channels span the greasy cell membrane of many cells, including those in your nerves and muscles, but this one is from a bacterium, which of course has neither nerves nor muscles. However, just like your potassium channels, the function of this particular ion channel is to let one, and only one, type of **positively charged ion** (an atom that has lost an electron) pass through from the watery soup inside the cell to the other side of the cell membrane, outside the cell. These potassium ions would never make it through the slimy membrane of the cell in a timely fashion otherwise, even though another protein has pumped the cell chock-full of them. The ions need a special passageway, and these protein channels make a gloriously inviting hallway for them. On one end is a door. When the door is open, the large chamber within the channel connects with the watery inside of the cell. On the other end is a narrow passageway that only potassium ions, not even smaller **sodium ions**, find inviting. In this way, the channel acts as a **filter**, letting only potassium ions and no appreciable amount of any other kind of ion pass through.

The goopy interior of the cell has a lot of water, each molecule of which has an **oxygen** atom on one side and two **hydrogen** atoms sticking out at an angle on the other side. Because there is a slight negative charge on the water's oxygen atom, which hogs a bigger portion of the electron cloud surrounding the water molecule, the positively charged potassium ion is attracted to the oxygen side of the water. In fact, the potassium ion manages to entice several water molecules to align their oxygen sides toward it. As the potassium ion moves through the water, it is literally handed off from water molecule to water molecule, just like the hands of spectators move a person balanced above them in a mosh pit. As a potassium ion approaches the narrowing in the channel, the part that acts as a potassium filter, the oxygen atoms hand off the ion to a series of tightly spaced oxygens in the filter. The potassium doesn't recognize that these oxygens are part of the channel's structure, contributed by specific amino acids and not by another water molecule. In this way, potassium ions are able to pass through the hallway to the other side of the membrane at the rate of 100 million per second, when the rear door is open. Given how many can pass through at one time, it is important that this channel's door can shut tight, to prevent too much potassium from bleeding out of the cell. The cell depends upon charge differences set up by ion pumps to move many chemicals across the membrane. If the doors to all ion channels were open all the time, the cell would exhaust the pumps and have a hard time maintaining a difference in charge between the inside and the outside of the cell.

Stop and Consider

Nucleic acids like DNA and mRNA have many negative charges on the outside to which positive ions found in the cytoplasm are drawn. How do you think mRNA might exit the nucleus? Explain your answer.

The importance of the precise structure and hence functioning of protein machines like these channels cannot be overstated. Potassium channels, like other channels that pass other ions from one side of the cell membrane to the other, have a particular architecture that allows them to open and close upon command. We now know that intricately designed and mechanically fine-tuned ion channels determine the rhythm of our heartbeats and allow an electrical impulse initiated when we stub our toe to be transmitted to our brains. We also know that defects in these channels can cause disease. The variations in our genes that code for the making of imperfect channels are often inherited from our parents. For example, long QT syndrome 1 is caused by a defect in a potassium channel encoded by a gene on our 11th chromosome. Q and T are letters on the electrocardiogram, the tracing of the electrical impulses of our beating hearts, and long QT means that a certain part takes a bit too long. The malfunctioning potassium channels of people with the long QT syndrome can cause cardiac arrhythmias and sometimes sudden death.

In addition to inherited DNA differences, we can also accumulate defects in the DNA code of a single cell. Many cancers originate this way. For example, the protein RB1 normally binds another protein called E2F and keeps it from directing production of the proteins that orchestrate DNA synthesis as cells prepare to divide. As long as RB1 can hold back E2F, DNA synthesis does not occur until the cell is ready for this to happen. However, a defect in RB1 can cause it to let go of E2F prematurely. Consequently, DNA synthesis occurs before the cell is ready. If a normal body cell acquires a particular change in the DNA that codes for RB1, the cell may divide over and over, released from the normal controls on cell division. Such unregulated cells can grow out of control to cause cancer. It is important to understand how the DNA code specifies the production of precision protein machines such as ion channels, RB1, and E2F.

HOW DOES DNA CODE FOR PROTEIN MACHINES?

After the structure of DNA was determined, the next great mystery was how the four kinds of bases found in the long molecules of DNA could code for the making of proteins using approximately 20 different kinds of amino acids. Crick himself had suggested that the code for one amino acid was composed of a sequence of three bases that scientists call **A**, **T**, **G**, and **C**. At any rate, Crick later suggested that one could read a protein code or gene *by following the bases down just one side of the DNA ladder.* This much turned out to be true so that depending upon where one starts reading down the side of the DNA ladder, the first three bases, such as "CCA," ultimately determine the first type of amino acid at the beginning of the protein coded for by this particular section the DNA ladder. The next three bases, down the side of the DNA ladder in order, *without skipping any bases*, such as "GCT" determine the second type of amino acid in the protein chain, the next three determine the third amino acid, and so on until one reaches a set of three bases such as "TAG" that will say "STOP." Each set of three bases is called a **codon**.

However, Crick erroneously thought that there was only one three-base code for each amino acid. It turns out that the 64 possible combinations of three bases (4 possible bases in the first position times 4 possible bases in the second position times 4 possible bases in the third position) all code for only around 20 types of amino acids, because several triplet codes or sets of three bases such as "CAA" or "CAG" code for the same type of amino acid. In addition, there are *three* triplet codes for "STOP," at which point addition of new amino acids stops (Figure 2.2).

Note that DNA resides in the nucleus of the cell, but that the proteins are assembled outside of the nucleus in another part of the cell. So how does the DNA code *in the nucleus* determine what proteins are made *outside of the nucleus*? Good question!

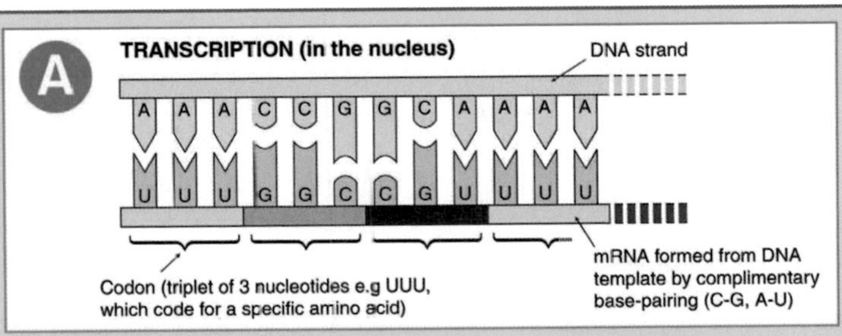

A

TRANSCRIPTION (in the nucleus)

DNA strand

A A A C C G G C A A A A

U U U G G C C G U U U U

Codon (triplet of 3 nucleotides e.g UUU, which code for a specific amino acid)

mRNA formed from DNA template by complimentary base-pairing (C-G, A-U)

B The Genetic Code

1st position (5' end)	2nd position U	C	A	G	3rd position (3' end)
U	Phe	Ser	Tyr	Cys	U
	Phe	Ser	Tyr	Cys	C
	Leu	Ser	STOP	STOP	A
	Leu	Ser	STOP	Trp	G
C	Leu	Pro	His	Arg	U
	Leu	Pro	His	Arg	C
	Leu	Pro	Gln	Arg	A
	Leu	Pro	Gln	Arg	G
A	Ile	Thr	Asn	Ser	U
	Ile	Thr	Asn	Ser	C
	Ile	Thr	Lys	Arg	A
	Met	Thr	Lys	Arg	G
G	Val	Ala	Asp	Gly	U
	Val	Ala	Asp	Gly	C
	Val	Ala	Glu	Gly	A
	Val	Ala	Glu	Gly	G

C Amino Acids and Their Symbols

A	Ala	Alanine
C	Cys	Cysteine
D	Asp	Aspartic acid
E	Glu	Glutamic acid
F	Phe	Phenylalanine
G	Gly	Glycine
H	His	Histidine
I	Ile	Isoleucine
K	Lys	Lysine
L	Leu	Leucine
M	Met	Methionine
N	Asn	Asparagine
P	Pro	Proline
Q	Gln	Glutamine
R	Arg	Arginine
S	Ser	Serine
T	Thr	Threonine
V	Val	Valine
W	Trp	Tryptophan
Y	Tyr	Tyrosine

Figure 2.2 During transcription, mRNA forms opposite a portion of one side of the DNA that corresponds to a gene (A). Note that mRNA follows the same pairing rules as DNA, except that the base U is substituted for the base T to form U-A pairs instead of T-A pairs. Each three bases along the mRNA ultimately determine which of the 20 amino acids will be incorporated into a protein. The mRNA code is shown in part B. For example, the sequence C (first base), A (second base), A (third base) codes for the amino acid glutamine, abbreviated Gln. Proteins are made up out of 20 kinds of amino acids. Their abbreviations, both three-letter and single letter, are shown in part C.

HOW DOES THE CODE GET OUT OF THE NUCLEUS?

To answer that question, we need to introduce another flavor of LifeSavers—*watermelon.* It turns out that kelp, kangaroos, and humans, as well as every other organism, make proteins in pretty much the same way. To solve the sticky (no pun intended) problem of how a DNA code in the nucleus determines which amino acids are put together to make protein chains outside of the nucleus, we must discuss another type of LifeSaver string called **messenger RNA** or **mRNA** for short. Instead of being composed of cherry, orange, lemon, and lime (the bases G, C, A, and T), it is made up of cherry, orange, lemon, and watermelon (the bases G, C, A, and **U**). In other words, mRNA uses the base U instead of the base T and makes U-A pairs instead of T-A pairs, when it has the opportunity to pair up with another string of bases. Instead of being made of two long strings of bases attached to two sugar-phosphate backbones like the DNA ladder, mRNA is made of only *one* string of bases attached to a sugar-phosphate backbone (i.e., is single-stranded) and is relatively short.

In order to copy the DNA code of one gene, a set of proteins help another protein called **RNA polymerase** open up a short portion of the double-stranded DNA. RNA polymerase then puts together a single-stranded mRNA using a section on one side of the DNA as a pattern or **template**. That section is a gene. The same pairing rules apply so that at least momentarily, the mRNA with its cherry, orange, lemon, and watermelon candies is paired up with the appropriate partners on this section of DNA. Note that mRNA's U pairs with DNA's A and otherwise all the other pairings are similar to the DNA matches: mRNA's G pairs with DNA's C, mRNA's A pair with DNA's T, and so on. When construction is finished, the completed single-stranded mRNA separates from the DNA and the DNA snaps back together again.

In bacterial cells, the mRNA is used as is to direct the assembly of proteins. In the cells of higher organisms, a chemical **cap** is added to the front end of the mRNA and a series of A's, called a **poly-A tail**, are added onto the far end of the mRNA (Figure 2.3). In addition, in cells of higher organisms (those with a nucleus) portions of the mRNA sequence called **introns** are removed and the remaining portions called **exons** are spliced back together before the mRNA exits the nucleus. Introns may contain sequences that regulate transcription of DNA to mRNA. The mRNA then leaves the nucleus to take the coded message to the protein factory within the same cell—hence the term "messenger RNA." There, large workbenches made of both protein and nucleic acid grab the mRNA so the correct amino acids can be brought up to the mRNA. Each amino acid is escorted by a molecule called **tRNA** or **transfer RNA** (refer again to Figure 1.3). It is important to note that the escort molecules have three bases prominently exposed on their backsides and that these molecules also use the base U instead of T. The kind of amino acid linked to the escort is determined precisely by the tRNA escort's **anticodon**, or triple set of bases on the escort's backside. Note that the tRNA for a different amino acid would have a different anticodon on its backside. Therefore when the tRNA escort momentarily attaches its anticodon to the triplet of bases in the mRNA, the tRNA brings into place its particular type of amino acid, which it attaches to the growing string of amino acids. In this way, amino acids are attached in a *precise order* to make a particular protein! Take a minute to study Figures 1.3 and 2.2 to understand this code, because it is the basis for everything else you will learn in this book.

Therefore, the DNA code in the nucleus, which determines the complementary mRNA code also in the nucleus, ultimately determines the precise order and length of the string of amino acids that make a protein machine in another part of the cell. Because there is a direct correspondence between the DNA code and the string of amino acids that make up a protein, the DNA determines the specific amino acid sequence of each type of protein.

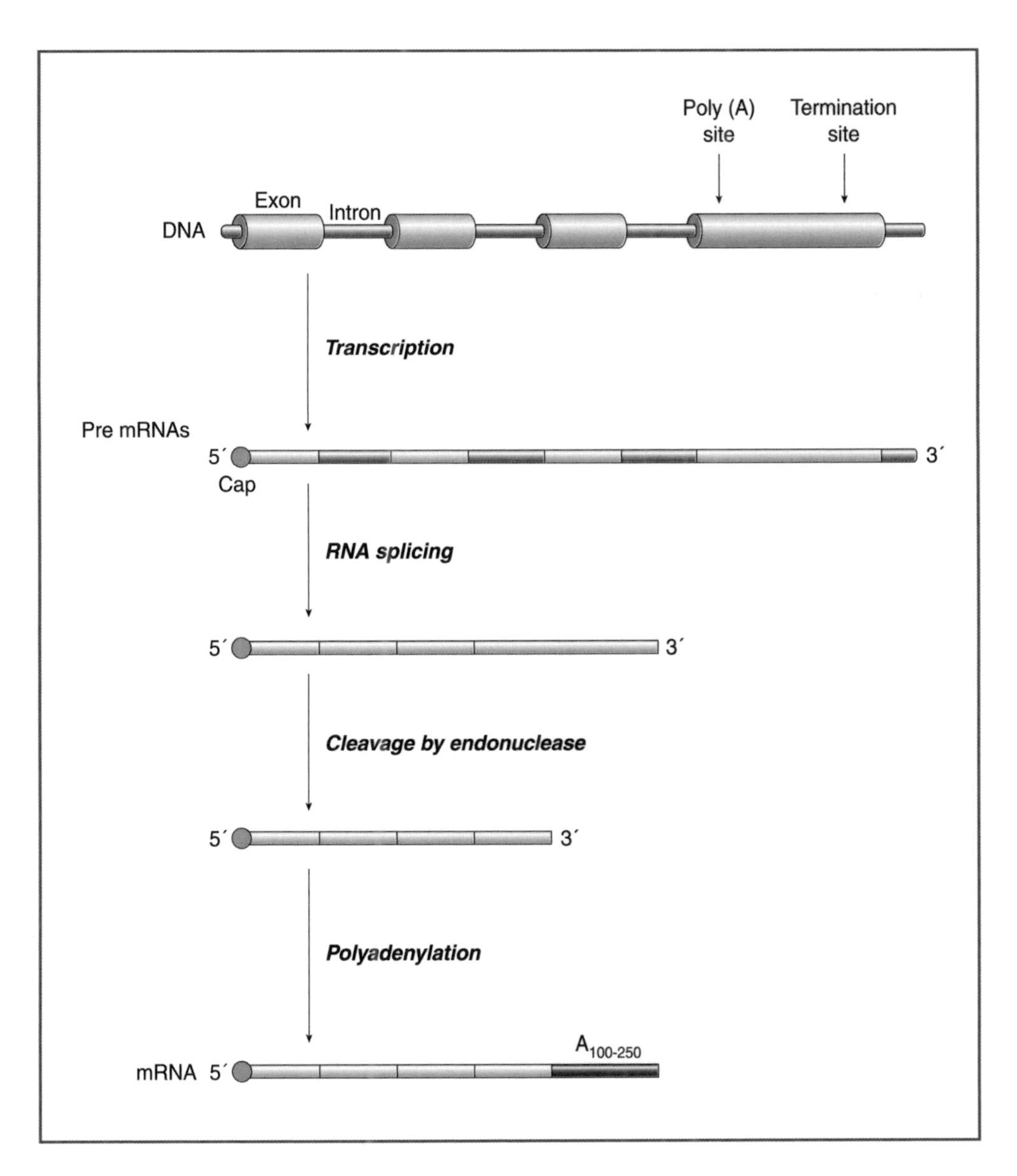

Figure 2.3 Before mRNA in most eukaryotes leaves the nucleus, any introns must be spliced (cut) out. The remaining exons are then spliced together. A chemical cap is added to the 5' end. After a piece of the 3' end that does not code for protein is cut off by an enzyme called an endonuclease, a poly-A tail (a string of A's) is added to the 3' end. Such a mature mRNA is called a **processed mRNA**.

PROTEINS ARE MADE OF A STRING OR SEVERAL STRINGS OF AMINO ACIDS

Proteins are very different from the DNA string of LifeSavers. Proteins are composed of approximately 20 different types of subunit amino acids, like 20 different kinds of beads from which one could choose to string together a necklace. Just like different kinds of beads, such as a Venetian glass or African ebony bead, each of the 20 kinds of amino acids has its own special characteristics. Each kind of amino acid has a unique side part that sticks out. This protrusion may be long, short, or shaped like a fat pillow. In addition, this side part might be greasy and therefore easily slide next to a nearby protein that has greasy amino acids or harbor a positive or negative charge that could make it happy to reside in the watery cytoplasm. (Remember that "unlike charges" attract and "like charges" repel each other). This special part of the amino acid bead may be relatively fixed in space or inclined to flip around. Depending upon the order in which these beads are strung along in a line and then precisely become intertwined as the string folds back upon itself, the protein created will have nooks and crannies that are floppy or not, filled in or open, and greasy or charged. Such is the fate of a protein. Because the string will be folded in only one possible way, the sequence of particular amino acids initially strung together in one long line of 100 to 1,000 beads determines the shape and mechanical properties of the final protein.

Proteins such as the potassium channel are made up of essentially 20 different kinds of amino acids. Figure 2.4 shows the general formula for an amino acid. It has two different **functional groups**—at one end an amine ($N–H_2$) functional group and at the other end a carboxyl ($–COOH$) functional group. The **side chain** or "R" group that is attached to the middle carbon gives each amino acid its special characteristics. There are 20 usual types of amino acids. For example, if R is only a hydrogen atom, the

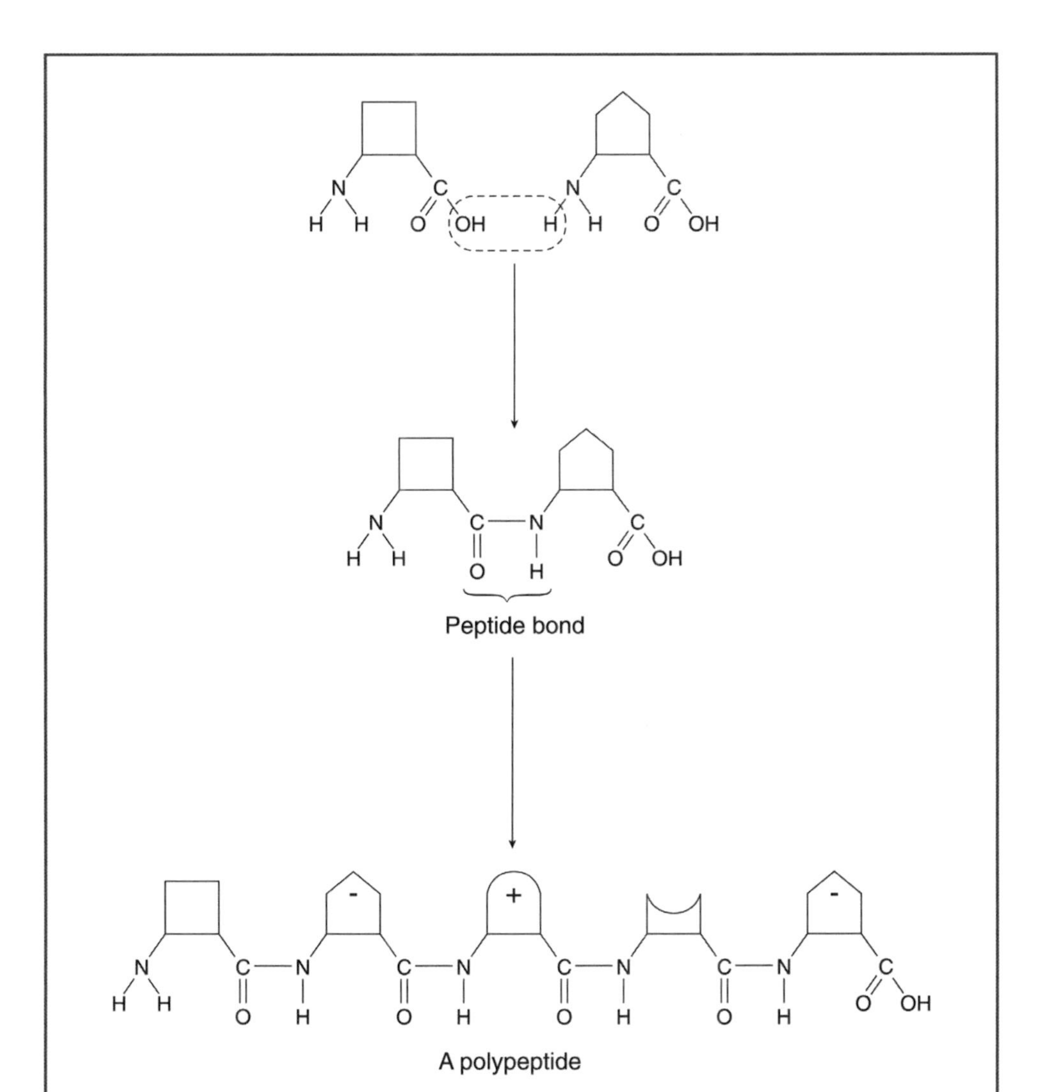

Figure 2.4 The basic structure of an amino acid is illustrated here. Each amino acid has an amine end (NH_2–) and a carboxyl end (–COOH), and a side chain attached to a central carbon. (The side chains and the central carbon are shown here as a geometric shape and any charge present on the side chain is represented as a "+" or "-".) A polypeptide, or protein, is a string of amino acids linked together through peptide bonds.

amino acid is **glycine** (abbreviated as G), but if R is a methyl group ($-CH_3$), the amino acid is **alanine** (A).

Amino acids are linked to one another like beads in a necklace when the amine end of one amino acid is joined to the carboxylic end of another amino acid. Linking several end to end creates a **polypeptide.** A protein can be made up of one polypeptide chain, or as we have seen for the potassium channel, several polypeptides folded together. In any polypeptide, the first amino acid, which has a free amine, is at the **N-terminus** of the chain and the last amino acid, which has a free carboxylic acid, is at the **C-terminus.**

The 20 amino acids can be divided into categories, depending upon whether or not their side chains are **hydrophobic** or **hydrophilic** (Figure 2.5). Hydrophilic side chains have a positive or a negative end just like a magnet. If the charge is weak, the side chain is said to have a partial charge, just like the greedy oxygen within a water molecule. On the other hand, if the side chain has captured an **electron** or a **proton**, it will have a net negative or positive charge. Such side chains can strongly attract oppositely charged cousins just like strong magnets. They can also push away molecules that have a similar charge. Hydrophobic amino acids, on the other hand, are greasy and frequently contain only carbon and hydrogen atoms. They are described as being hydrophobic (literally, fearing water), given their reluctance to extend an atomic hand to the partially charged ends of water to create hydrogen bonds. For example, oils are hydrophobic. You see evidence of the hydrophobic effect every day when the oily part of a salad dressing segregates itself toward the top of the bottle as it sits in the refrigerator. In contrast, many of the polar side chains of the hydrophilic amino acids are happy to create hydrogen bonds between themselves and water or other polar molecules.

Although we often categorize amino acids in this way, the hydrophobicity (or hydrophilicity, the affection for water) of their

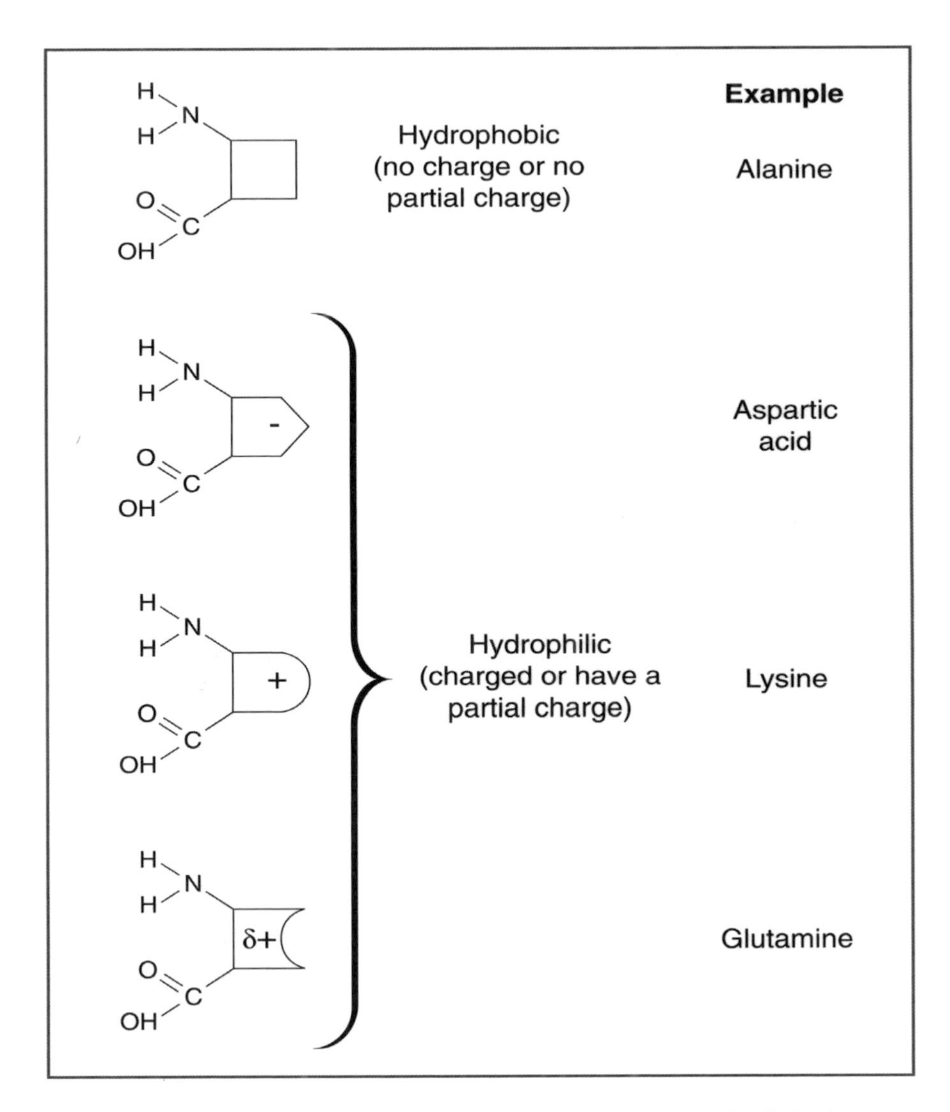

Figure 2.5 Amino acids are distinguished by the side chain linked to their central carbon. Here the kind of side chain and its link to a central carbon is represented by a geometric shape. Hydrophobic amino acids that like to avoid water have a greasy side chain made of mostly carbons and hydrogens. Hydrophilic amino acids that like to form weak hydrogen bonds with water have a charge (+ or –) or partial charge (δ+ or δ–) on their side chains that include nitrogen or oxygen.

side chains is a continuum that ranges from charged to incredibly greasy. Only five amino acids have charged side chains in cells: **lysine** (K) and **arginine** (R) have positively charged side chains, while **glutamic acid** (E) and **aspartic acid** (D) have negatively charged side chains. Only sometimes does **histidine** (H) have a positively charged side chain.

Perhaps now you can see how a change in the DNA code, a **mutation**, may change the amino acid incorporated, and therefore change the length of the protrusion, or change a positive charge to a negative one, or interrupt a greasy façade by inserting the wrong bead in the protein, which, of course, might throw a monkey wrench into that protein machine! That penguins, gorillas, humans, water striders, *E. coli*, and house plants all use essentially the same code and the same processes by which sections of DNA are periodically transcribed into mRNAs and then translated into particular proteins is remarkable. What's even more remarkable is that several of the most important sequences, such as the amino acids that line the narrowest part of the potassium channel have been preserved over hundreds of millions of years of evolution! Although the potassium channel shown in Figure 2.1 is from a bacterium, essentially the same design is used for the **selectivity filter** portion of human potassium channels.

Stop and Consider

Do you think that through a random mutation it is easier to make a defective protein that does not work or one that becomes overly active? Why?

HOW IS GENOMICS CHANGING THE WAY GENES AND PROTEINS ARE STUDIED?

The modern era of genomics, in which we have learned the

complete DNA sequence of each chromosome of many types of living things, has reinforced the interrelatedness of all living organisms, even down to the common structures of proteins like those that are found in organisms from bacteria to man. It is because of this commonality that we have learned so much about living systems by studying bacteria, yeast, plants, flies, and worms through **genetic analysis**. For years, scientists have tried to determine the genes responsible for a particular protein by altering the normal DNA sequence in living things and seeing what abnormality results.

For example, there are flies whose legs shake whenever they come in contact with **ether**, a foul-smelling chemical that was used to put children to sleep when they had their tonsils out 50 years ago. Scientists gave two of the mutations that cause this behavior the colorful names "Shaker" and "Ether-a-go-go." When scientists compared the DNA sequences in the chromosomes of shaking and normal flies, they were able to pinpoint the causes for that behavior in the alteration of two particular inherited DNA sequences. The error for each was found within the sequence that codes for two potassium channels. Evidently, the change in the structure of the

Using the Internet to Understand Transcription and Translation

If you have access to a computer with an Internet connection, visit *www.biointeractive.org* to view wonderful animations in real time of the process by which the DNA code is converted to an mRNA code (transcription) and the process by which the mRNA code is used to make a protein (translation). At the site, click the "DNAinteractive" icon. Then click on the tab "Code" and then on the tab "Using the Code." At each stage you will need to patiently wait to allow the movies to load. In order to view a transcription animation, click on the tab "Putting it Together" and then on the "Transcription" icon. To view an animation of translation, use the following path: *www.biointeractive.org* – DNAinteractive (icon)–Code (tab)–Reading the Code (tab)–Putting it together (tab)–Translation (icon).

channel coded for by the altered DNA blueprint was causing the strange behavior of these flies. By studying the effects of changes in the DNA blueprint of fly potassium channels, by inducing the same changes in worm and mouse potassium channels, and by pinpointing the DNA errors in the codes for human potassium channels and proteins that regulate them, we have learned a great deal about heart and muscle diseases.

Long QT syndrome 1, as described above, can cause cardiac arrhythmias and sometimes sudden death and is caused by a mutation in a potassium channel gene on the 11th human chromosome. But similar heart problems can be induced by drugs, which cause many people to not be able to take prescription medications they otherwise would. Scientists found that the potassium channel gene in these patients was not mutated. Therefore, they began to look for other genes that might regulate the potassium channel. In a recent study, they used worms that could no longer thrash about as fast as usual (because of a mutation in their "ether-a-go-go" potassium channel) to find proteins that modified that channel's activity. Once they identified the worm genes that caused the same slowdown in thrashing when mutated, they checked patients suffering from the drug-induced long QT syndrome. Sure enough, they found that these patients had mutations in the human genes for the regulatory proteins for the potassium channel. This finding suggests that defects in regulatory proteins that made these people susceptible to heart problems when taking prescription drugs were causing their "induced" long QT syndrome.[3]

Intrigued by these common protein structures and the commonality of the underlying DNA code, scientists now want to mine the data of over a thousand species of living things whose DNA we have completely decoded, not only to learn about human disease but to learn more about the inner workings of living things and their evolutionary relationships to one another. For example, we know the order of all 3 billion bases in human chromosomes. Similarly, we have decoded 120 million of the 160 million bases of the fly genome, all 125 million bases of the green plant *Arabidopsis* genome, the 100 million letters of the worm *C. elegan*'s DNA, all 12 million letters of the yeast genome, and all 5 million letters of the bacterium *E. coli.* Therefore, not only can we compare genes that occur in each of these organisms, but we can also compare their positioning along the DNA in the chromosome, a helpful technique given the fact that some genes are found in the same order in spots on several different genomes. For example, the genes in many sections of mouse chromosomes are not only similar to human genes, they are present in the same order along the chromosome.

Biologists have been studying genes one at a time since Watson and Crick introduced the structure of DNA and their colleagues deciphered the correspondence of the three-letter code for a particular amino acid. In contrast to the analysis of one gene at a time, in which it took years for a laboratory to figure out the code of a gene, determine under what conditions a protein corresponding to this gene was made, and determine the protein's function, this new age of genomics has inspired the development of ways to study all of the mRNAs or all of the proteins made by cells all at one time. Because the mRNA profile may differ from one type of tumor to another or between diseased and normal tissue, such new **high-throughput** methods are becoming useful diagnostic tools.

Researchers have also worked to speed up the laborious process of determining the 3-D molecular structures of proteins. Because we know the 3-D structures for more and more proteins, bioinformatics specialists have developed computer tools to predict an unknown protein's 3-D structure based upon a gene's DNA sequence. The programs combine the 3-D structure information obtained the conventional way using x-ray analysis of protein crystals with the DNA sequence for the protein to develop rules to predict the 3-D structure of a new protein based on its DNA sequence. What is amazing is that this analysis is free and available to anyone who just points their Web browser to the appropriate Website, such as the National Center for Biotechnology Information (NCBI) (at http://www.ncbi.nlm.nih.gov), which you will explore later in this book. The magazines or journals that scientists use report all of this information about genes, mRNAs, proteins, mutations, and the effects of these gene changes upon protein structure and the behavior of cells, **tissues** made up of a particular type of cell, and entire organisms. The articles may also discuss the interactions of various proteins that affect the operation of their targets. Links to abstracts or summaries of this work are also searchable at the NCBI server.

PRO OR CON?

Because of genetic testing, we are now able to determine many defects present in a fetus before the baby is born. In some cases, this will lead the expectant mother to alter her diet or in some other way provide the best possible environment for the growing fetus. In other cases, the parents might consider terminating the pregnancy. Would you want to have such tests done early in the pregnancy? Why or why not? Are there any circumstances under which you might have such testing done?

CONNECTIONS

Proteins like the potassium ion channel are precision machines. They are made up of either one or several long chains of amino acids folded together. The particular sequence of amino acids is critical in determining a protein's structure and function.

DNA in the nucleus codes for making proteins outside of the nucleus. The code must be transcribed into another molecule called mRNA that can travel out of the nucleus. The enzyme RNA polymerase makes mRNA by using a portion of one side of the DNA molecule as a template. In cells more complex than bacteria, the mRNA is capped and the introns spliced out, and a poly-A tail is added before the mRNA leaves the nucleus.

The mRNA is translated into a particular protein chain outside of the nucleus. A workbench made of both protein and RNA grabs each mRNA molecule and holds it so that other RNA molecules called transfer RNA can bring particular amino acids into place to be snapped together like beads in a necklace to make a protein chain. Each triplet code in the mRNA pairs with the anticodon of a tRNA using the U-A and G-C coding rules. A different kind of amino acid is brought up depending upon the type of tRNA. Because there are 64 possible three-letter codes for the approximately 20 amino acids, there is more than one triplet code that specifies each amino acid.

Each amino acid has particular properties and these help to determine the characteristics of the protein machine. Amino acids are linked end-to-end to make a protein chain and each protein is made of one or more chains folded together. The properties of the protein machine are determined by the order of amino acids because the side-chain of each amino acid has specific properties. Only five of the usual 20 amino acids can have a positive or negative charge. Mutations or changes in the DNA code will result in defects in the protein machine made because the wrong amino acid might be incorporated at that spot.

Genomics, proteomics, and bioinformatics are changing the way the study of biology is conducted. High-throughput technologies allow us to determine all of the mRNAs and sometimes all of the proteins in a cell. Bioinformatics allows us to access molecular biology information deposited in databases throughout the world. By studying the genomes of a variety of organisms, we can learn more about our interrelatedness and about specific proteins preserved through time by evolution.

FOR MORE INFORMATION

For more information about the concepts discussed in this chapter, search the Web for the following keywords:

Potassium ion channel, Long QT syndrome, Triplet code, mRNA, tRNA, Genomics

3

The Human Genome Project

"We felt that the human genome sequence is the common heritage of all humanity and the work should transcend national boundaries," said Dr. Eric Lander of the Center for Genome Research at the Whitehead Institute for Biomedical Research. He spoke for more than 200 coauthors expressing their reasons for a collaborative worldwide effort to determine the base-by-base sequence of the human genome.[1] This group, called the **International Human Genome Sequencing Consortium** (**IHGSC**), was in direct competition with Dr. J. Craig Venter, founder and head of Celera Corporation. For Dr. Venter, the human genome was a goldmine whose sequences could be mined for profit. Remarkably, the competition spurred both groups to produce an enormous amount of sequence at a blistering pace; the competition ended in a virtual tie. Their largely consistent drafts of the human genome were unveiled for the public in February of 2001. The International Consortium published their results in the highly respected scientific journal *Nature*.[2] Dr. Venter and his collaborators in the United States, Israel, and Spain published theirs in the equally esteemed journal *Science*.[3]

Perhaps we see ourselves differently, knowing the complete genomes of over a thousand species. The genes we share with other species, sometimes with an almost exact match for the encoded protein, are humbling. In 1866, the Augustinian monk Gregor Mendel announced that traits are inherited in discrete packets. He had just spent seven years in the monastery's garden cross-pollinating pea plants (*Pisum sativum*) and recording the characteristics of their offspring. That momentous conclusion, ignored during Mendel's lifetime, was rescued from obscurity at the dawn of the 20th century. The intervening hundred years have cemented the relationship between those discrete packets, or genes as we call them, and the traits they influence. Years of painstaking and often repetitive work by scientists and their students have defined gene sequences in a variety of model genetic organisms, paving the way for the high-throughput technology and the international cooperation that made sequencing whole genomes possible.

Each success in science builds upon past successes. The opportunity to push the frontiers of knowledge requires not only individuals with vision and nerve, but advances in technology that allow their dreams to be realized. The effort to sequence the entire human genome was not unique, except perhaps in terms of scale and the degree of cooperation that would be required of scientists around the world. The public effort involved 20 groups from the United States, the United Kingdom, Japan, France, Germany, and China and considerable governmental and private funding. The early successes that made sequencing the human genome seem even remotely possible were the complete sequencing of several small viruses and the human mitochondrion, the energy factory in a cell, by the early 1980s. Dr. Lander said: "These projects proved the feasibility of assembling small sequence fragments into complete genomes, and showed the value of complete catalogues of genes and other functional elements."[4]

AN IDEA TAKES HOLD

The idea of sequencing the entire human genome began to take form at meetings in the United States between 1984 and 1986. A report by the U.S. National Research Council in 1988 proposed not only generating a map of the human genome, but also producing similar maps for model organisms including the genetic

Gregor Mendel, the Father of Genetics

Johann Mendel was a poor boy whose family sent him to Catholic school and encouraged him to join the priesthood. In 1843, he entered an Augustinian monastery in Brüun (now the major city Brno in the Czech Republic) at the age of 21 and took the monastic name, Gregor. Because of his interest in the natural sciences, he was allowed to attend Vienna University and returned to Brüun to teach school, at which he was not successful. But he also carried out experiments in the small garden next to the church of St. Thomas. His success with his experiments made him, eventually, the Father of Genetics. Between 1856 and 1863, he and his two helpers generated over 33,500 pea plants. He was attempting to see whether traits such as the color of the flower (purple or white), the color of the seed (yellow or green), the shape of the seed (round or wrinkled), or the length of the stem (standard or dwarf) could be passed on to the next generation in a predictable way. In other words, he asked this kind of question: "If I transfer pollen (the male **gamete** comparable to a sperm) from a purple plant to the flower of a white plant in order to fertilize the ovules there (the female gametes comparable to eggs), will the seeds produced create plants that are all purple, all white, or some combination of purple and white?" Mendel's answers to questions such as this helped him to formulate the idea demonstrated above, that traits of living things are inherited in discrete packets, one of each type inherited from the father and one from the mother. We call those packets genes and now know that they correspond to a segment of the DNA along a chromosome. (That molecular definition of a gene was not determined until the middle of the next century.)

workhorses: bacteria, yeast, flies, worms, and mice. At the inauguration of the **Human Genome Project** (**HGP**) in 1990, the United States, Great Britain, France, and Japan had genome centers. Other countries soon joined in so that by late 1990, the Human Genome Organization was founded to encourage international cooperation in sequencing the human genome.

Early projects tackled the yeast, worm, and human genomes on a piecemeal basis, including the sequencing and cataloguing of **ESTs** or **expressed sequence tags**. ESTs are sequenced pieces of **cDNA**, DNA

Mendel also asked whether three traits present in each parent would be inherited independently of one another or whether or not there would be a disproportionate number of offspring that looked like the parents. He moved pollen from plants that had purple flowers and round, green seeds to the flowers of plants that had white flowers and wrinkled, yellow seeds. Then he waited for the plants to produce seeds and recorded the traits of the new generation produced when these seeds grew into mature plants. His results established the Law of Independent Assortment, the idea that two or three traits are inherited independently of one another. Notice that this is not entirely true. He was lucky not to choose genes next to each other on the same chromosome because such linked genes are usually inherited together. However, because portions of the parental chromosomes are regularly exchanged through crossing over, genes that are far enough apart or on different chromosomes are in fact inherited independently. He also must have consciously chosen traits that were determined by only one set of genes, one inherited from the male parent and one from the female parent. In fact, many traits such as height in humans are the result of several sets of genes acting together to generate the observable trait.

Although Mendel presented his work at a meeting of the Brno Natural Sciences Society and published his findings in its journal (in German) in 1866, his findings were misunderstood and ignored during his lifetime. Not until the dawn of the 20th century were his findings rediscovered. They are the foundation of all genetic studies that have followed.

derived by producing a complementary sequence from mRNA from a cell. They provide a quick indication that a protein product is likely being made and that the corresponding gene is actually used by the cell. The nice thing about mRNA sequences is that they eliminate the necessity of pasting together fragments of genomic sequence. As described in Chapter 2, the complete sequence for one gene in eukaryotes is usually not contiguous on the chromosome. The protein coding sequence is interrupted by introns that cells faithfully include in the first mRNA copy of the gene, but then naturally cut out of that mRNA in a process called splicing. The processed mRNA that leaves the nucleus therefore has no introns.

Some Key Events Leading up to Sequencing the Human Genome

1977	First gene sequenced
1982	Bacteriophage lambda genome sequenced
1984–1986	Sequencing project discussed at scientific meetings
1986	Fluorescent tags used for sequencing
1990	Human Genome Project officially begun
1995	*Haemophilus influenza* bacterial genome sequenced using whole shotgun method
1996	A gene map of the Human Genome (tying certain gene sequences to locations on chromosomes) created
2001	Drafts of the Human Genome completed, covered ~90% of the euchromatic region with 1/10,000 base accuracy
2003	Finished Human Genome which covered ~99% of the euchromatic region with 1/100,000 base accuracy

The public and private funding to sequence the human genome was created with the expectation that knowledge of the sequences that code for proteins and the sequences surrounding them that help to regulate their use would be medically useful. It was anticipated that an annotated human genome would catalogue not only the causes of diseases from cancer to mental illness, but would also allow for specific diagnoses and designer drug therapies that target the product of a particular gene. Previously it had taken years to decode a single human gene, given the relatively sparse and scattered amount of protein coding sequence in a sea of introns. Although the average human gene is about 27,000 bases or nucleotides long, the portion that actually codes for protein sequence is only about 1/24th of that length, about 1100 bp (base-pairs) long.[5] It was also widely held that knowledge of the mouse and human genomes would allow modern medicine to further capitalize on their physiological, metabolic, and behavioral similarities. The expected **conserved segments of synteny**, large neighborhoods of genes preserved during evolution in the same order along sections of human chromosomes and their mouse counterparts, heightened the sense of the potential usefulness of completing both the human and mouse genomes.

HIERARCHICAL SEQUENCING

The public effort adopted a **hierarchical** or **top-down** method for sequencing the genome (Figure 3.1). Human DNA from at least eight different individuals was first cut into relatively large pieces about 150,000 bp long in a way that provided plenty of overlapping fragments. This was achieved by cutting the human genome with a restriction endonuclease (also called a **restriction enzyme**) such as *Eco*RI. Restriction endonucleases are powerful tools because they cut DNA only when they encounter a specific series of bases.

Restriction endonucleases have been isolated from bacteria, which use them to protect themselves by cutting up any foreign

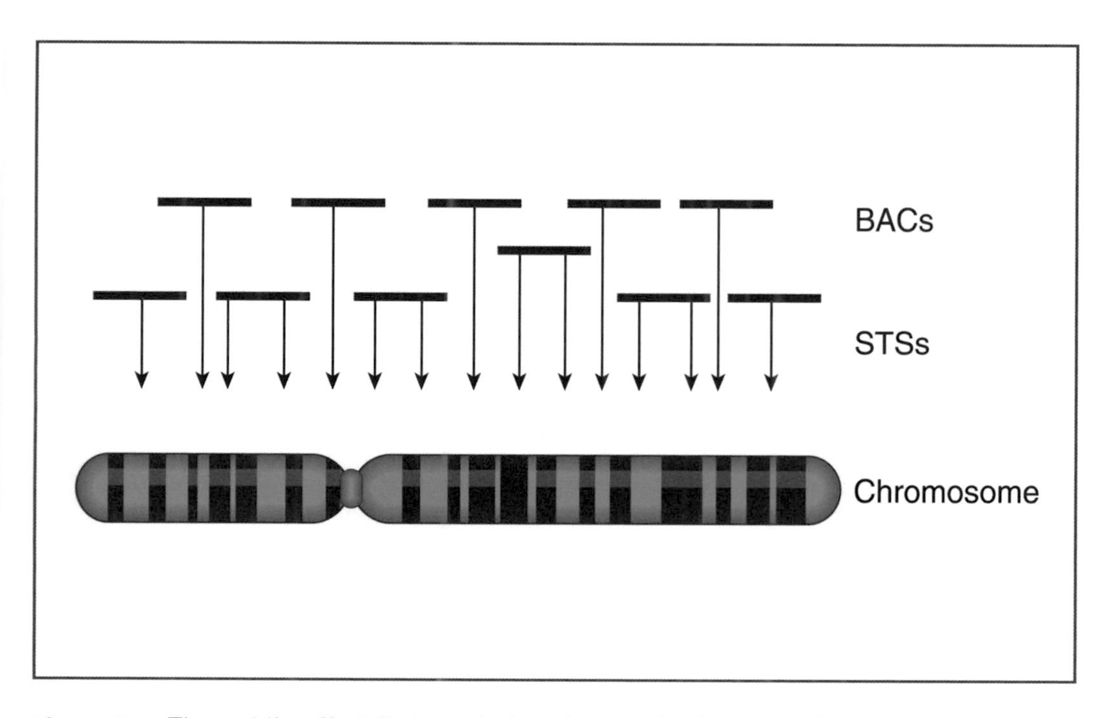

Figure 3.1 The public effort first used clone-based physical mapping to determine the order of a series of bacterial artificial chromosomes, or BACs, across each chromosome in a process called hierarchical sequencing. Then the researchers used known sequences called STSs to tie the BACs to specific locations on each chromosome. Finally, the BACs were shipped to labs around the world for sequencing by the whole shotgun method.

DNA that enters the cell. *Eco*RI looks for the sequence GAATTC along either side of the DNA helix. Note that because of the G-C and A-T pairing rules, this sequence will read GAATTC in both directions along the two backbones of a DNA molecule. Because *Eco*RI cuts between the G and the A on both sides of the DNA, it makes a jagged cut and generates an overhang on each end of the pieces of DNA. Such sequence-specific overhangs can be used to move pieces of DNA from one molecule to another by cutting both the donating and the receiving DNA with the same restriction enzyme. The public effort's first step in sequencing the genome was to move over a million pieces of cut DNA into another piece of DNA that bacteria would reproduce every time they divided. These extremely large pieces of DNA replicated inside of bacteria as bacterial artificial chromosomes or **BACs** (pronounced "backs"). Therefore, by growing large numbers of the bacteria in a simple broth of salts and nutrients, scientists were able to generate large numbers of identical copies of these pieces of DNA. This process of moving DNA and having bacteria make many copies of it is called cloning and the piece of DNA is called a **clone**. (Unfortunately, scientists have confused this term by also using it to describe the production of a genetic copy of a cell and even an entire individual, like Dolly the sheep!) How many of these BACs would be required minimally to "cover" a sequence that is 3 billion bases long? Hint: to find the answer divide 3,000,000,000 bp by 150,000 bp, the average size of a BAC insert. Although only 20,000 clones of 150,000 bp each are required to extend the entire length of the human genome, scientists cloned 65 times that amount so that there would be plenty of overlapping clones, places where the end of one clone included sequences that were present in the beginning of another clone. They also cut the many copies of human DNA for a relatively short time so that not all the *Eco*RI sites were cut in each copy of the genome. This kind of incomplete cutting created large (150,000 bp segments on average) and overlapping clones. They also created other libraries of clones with overlapping sequences by cutting these sets with different restriction enzymes.

For the public effort, the first challenge was to arrange some of these pieces of BAC DNA in the order they appeared along each chromosome. The first goal was to determine a relatively minimally overlapping set of BACs that would correspond to the DNA sequences in each of our 22 types of **autosomes** (non-sex chromosomes) and in an X and a Y chromosome. Almost all of the human DNA used in the sequencing project came from men. The major way in which the public effort to sequence the human genome differed from the provate one was this first step, **clone-based physical mapping,** which is described below.

An inexpensive way to look for overlapping regions was to digest each BAC with the restriction endonuclease *Hin*dIII that would cut each BAC into smaller pieces. Then the pieces from each digested BAC were sorted by size. The objective was to compare the sizes of pieces generated from different BAC clones and use this information to see which BAC clones might overlap because they contained at least some pieces that were the same size (Figure 3.2). Pretty clever! Actually this kind of clone-based physical mapping has often been used to determine the order of pieces of DNA. But the lengths of DNA whose order was determined by such mapping in the past had been much shorter than 3 billion base pairs long! To accommodate the complexity of the comparisons required for putting in order all of the pieces generated from clones from the human genome, computers were enlisted to read the sizes and decide which clones had overlapping ends.

To determine the sizes of each piece of DNA generated by digesting a BAC, the following steps were taken: The pieces of DNA were sorted by size in electrophoresis gels (Figure 3.3). A sample of the contents of the pot of pieces from each digested BAC was loaded into a well at the end of a gelatinous slab immersed in a salty liquid. Many such samples could be loaded at the same time, each in its own well at the front of the slab. Then a power source was attached to both ends of the container so that an electric field was applied across the gel. The negatively charged DNA pieces were pulled through this gelatinous forest toward the positive end of the box. You can see that the small pieces would be able to move faster than the larger ones, just as you would be able to run through a forest faster alone rather than if you were linked arm-in-arm with 20 of your friends. After the DNA pieces were separated (and before the smallest pieces ran off the end of the gel!), the power

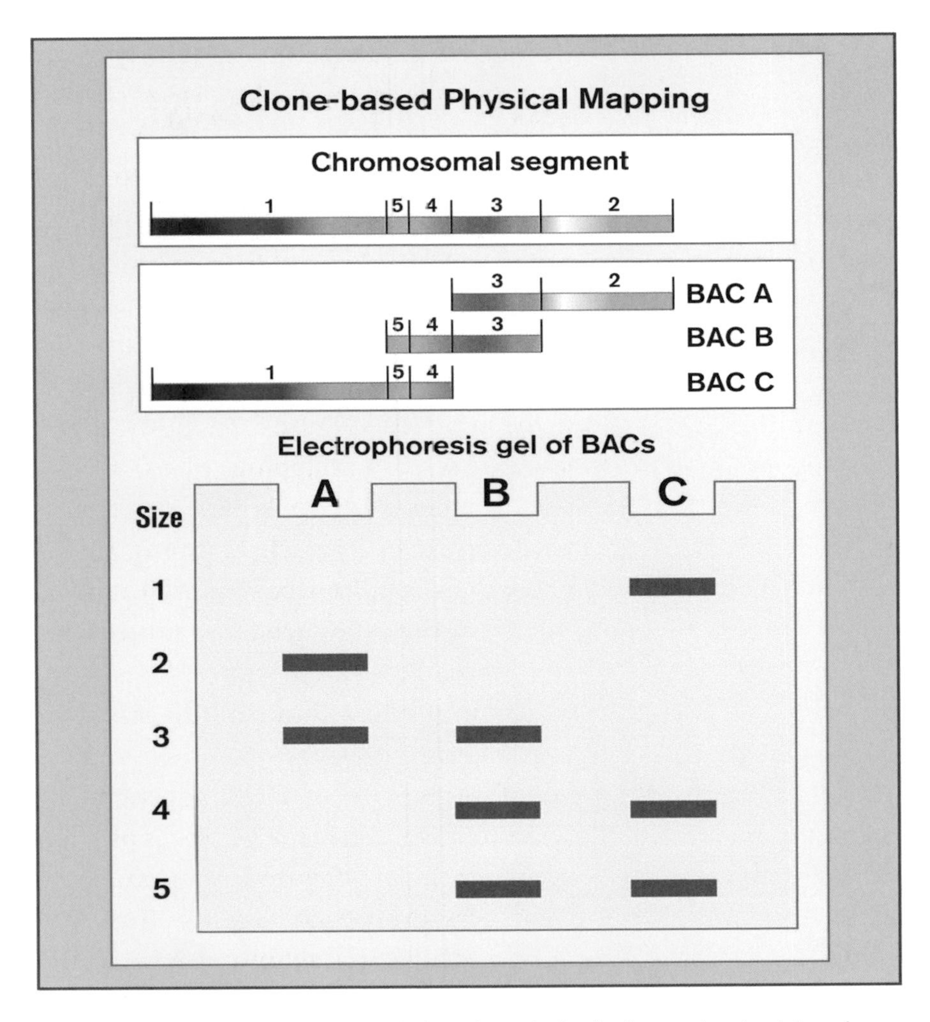

Figure 3.2 The public group used clone-based physical mapping to determine the order of BAC clones along the chromosomes. The BACs were all cut with the same enzyme. If the same stretch of DNA were present in two clones, each clone would generate at least one piece of cut DNA that was the same length. In this example (top), notice that BACs A and B contain DNA piece 3, and BACs B and C contain DNA pieces 4 and 5. By looking for clones that generated at least some of the same-sized pieces after running the cut DNA on an electrophoresis gel (bottom), researchers determined the order in which the clones should be aligned to cover the DNA corresponding to each chromosome.

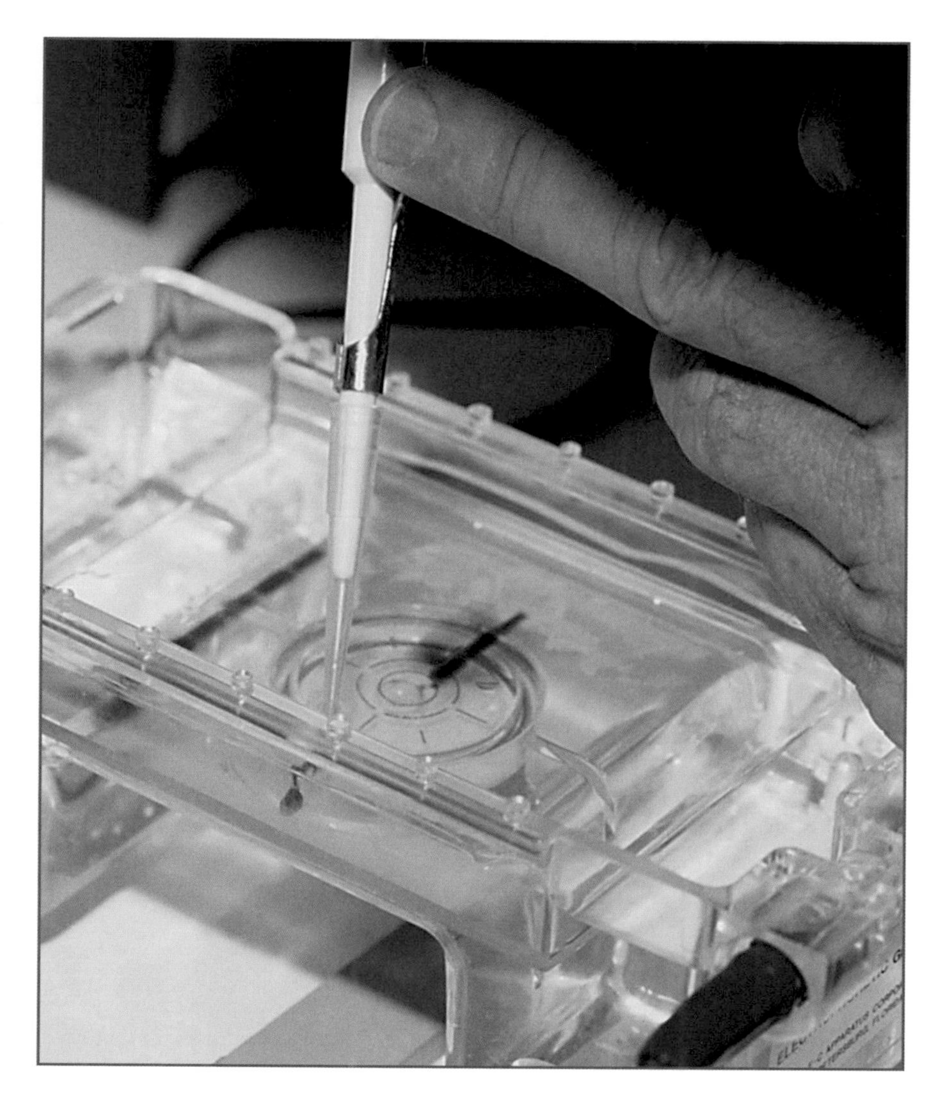

Figure 3.3 Gel electrophoresis is used to determine the size of pieces of DNA. After the BACs shown on the previous figure were cut with a restriction enzyme, the DNA was loaded into tiny wells within a gel slab in a box surrounded by liquid. DNA is negatively charged, and thus the DNA is pulled by an applied electrical field to the end of the gel slab. Because the smallest pieces run the fastest, the pieces are separated by size. An actual electrophoresis apparatus is shown here.

source was disconnected, and the DNA pieces were made visible with a special dye that shows up under UV light. On the picture of a gel, the DNA fragments appear as a ladder of lines, or a DNA "fingerprint," just like the DNA fingerprints that are used to solve crimes. In this case the scientists were looking for evidence from the gel that at least some of the pieces were the same sizes in different BACs. It is important to note that we are able to see the band of DNA because we have cut many identical molecules of DNA isolated from the many bacteria that created them. A single molecule of DNA would not show up.

Fingerprints made from BAC clones that share a good portion of their pieces that are the same size indicate that these BAC clones contain parts of each other. Because they contain parts of each other, stretches of their sequences are likely the same, perhaps on the right end of one and the left end of the other, with a third BAC having sequence like the right end of the first. Therefore, they overlap and should be arranged next to each other. Each group of overlapping BACs that can be used to generate a continuous piece of DNA sequence is called a **contig**. This word comes from the word *contiguous*, which refers to how the sequences from these BACs are next to each other on the chromosome.

Although this process indicated many long contigs, it did not connect the contigs to specific locations on chromosomes or even to particular chromosomes. The public sequencing group used several methods to tie the contigs to particular locations along the chromosome. Such a map is called a physical map of a chromosome.

Throughout the 1990s, the genome project scientists were able to locate short sequences of a few hundred base pairs or **STSs** (**sequence tagged sites**) to certain positions along each chromosome. How did the scientists do this? One method called **radiation hybrid mapping** required fusing pieces of human DNA (that had been irradiated) to hamster DNA inside hamster cells. Then scientists used **PCR**, also known as the **polymerase chain reaction**, a technique which is described below, to see if certain short DNA STS sequences were present. Because they had other ways to tell the order of these pieces along the intact human chromosome, the human genome scientists were able to determine if the sequences were on a particular section of a human chromosome. Another technique allowed them to match the STS sequence directly to the intact human chromosome by essentially putting a colored tag on a sequence derived from the STS sequence and seeing where the tag stuck to the chromosome.

The sequences of many of these STS tags were taken from the sequences of the ESTs that had been generated by making DNA copies using human mRNAs as templates. There are several ways to show that a sequence is present within a certain piece of DNA. One is to use PCR. If you know the sequences flanking or situated on both sides of any sequence, you can buy or make a single-stranded string of those flanking sequences about 30–50 bases long. These short strings or primers are used along with nucleoside triphosphates by an enzyme called a DNA polymerase to make many copies of the DNA sequence between the primers. But this polymerase is not just any DNA polymerase. It was isolated from organisms that live in hot springs and so the DNA polymerase can function even when it is very hot. An eccentric character in the annals of science is Kary Mullis, who was awarded the Nobel Prize in chemistry in 1993 for thinking up PCR and showing it could work.

But let's get back to our question. How did the scientists tie the STSs to a particular location along the chromosome? To answer that question, you must understand how PCR works and how it can indicate that an STS sequence is also on a particular chromosome. PCR works by using a heating and cooling cycle that is repeated every few minutes. During this process, the heat-resistant polymerase makes many copies of the DNA sequence between and including the primers. The hydrogen bonds between the bases in DNA are easily broken when the temperature is raised. Therefore,

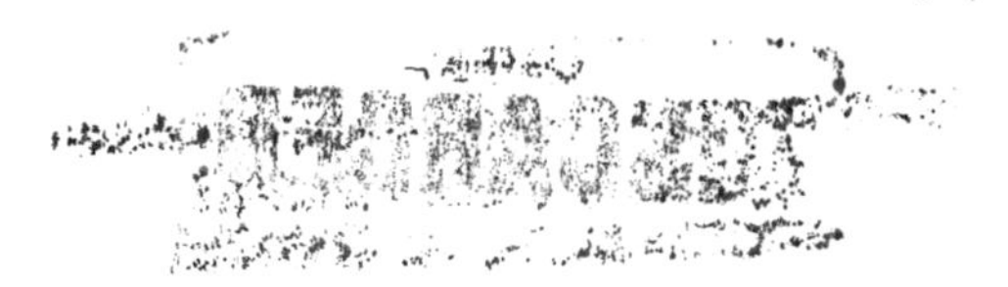

by raising the temperature to 95°C, the two sides or strands of DNA separate. After the strands have separated, the temperature is lowered to allow the primers to find their place on these strands. They will bind the DNA sequence that is complementary to their sequence. Both the primer and the DNA are single-stranded, except where they are bound to each other. Then the temperature is raised to an intermediate temperature so that the polymerase will add the A, G, C, and T nucleotides (originally as nucleoside triphosphates) to the primer, but only in the 5' to 3' direction (*only* on the sugar end of the previous nucleotide), and create a new piece of DNA using the **complementary strand** as a pattern. When the temperature is raised again, the strands separate. Then the cycle repeats itself. Pretty soon, the pieces of DNA that are as long as the primers plus the sequence on the DNA between them predominate in the mixture. At each cycle, the number of pieces doubles. Therefore, if you put the DNA polymerase, primers, nucleoside triphosphates, magnesium, extra ATP, a buffer to control the pH, and some DNA template into a small plastic tube, put the tube into a PCR machine for a few hours, and then load the contents of the tube into a gel for electrophoresis, you will be able see if a band of the expected size, the size of the primers plus the run of DNA between them, is there. If so, then your sequence was present on the DNA. Scientists prepared primers that corresponded to the sequences of STSs. Therefore, they could add the DNA from a radiation hybrid cell line plus the primers to a PCR reaction and test the results by electrophoresis. In this way, they could tell if the STS sequence was present within the hybrid cells. As previously mentioned, scientists could now determine where the STS was located along the human chromosome because they had other ways to tell the order of these pieces along the intact human chromosome.

Now can you imagine how scientists might check to see if an STS sequence were present in a BAC? During the Human Genome

Project, scientists often used PCR to check to see if a particular STS sequence was present in a BAC. Therefore, they could position the BAC and the entire contig it was a part of along the chromosome. But this was not the method they used most often.

Scientists used **probe hybridization** as the first way to tell if an STS sequence was present in a BAC and in this way determine the position of the BAC along the chromosome. This process involves making a probe or single-strand of DNA that will stick to its complementary strand. Because each strand of any DNA molecule will stick to its opposite or complementary strand, one can use a tagged DNA molecule as a probe to point out another DNA molecule that has at least a portion of the same sequence. A probe that corresponds to part of an STS sequence can be made radioactive by including nucleotides with radioactive atoms. Using some clever tricks, the probe can be used to find the same sequence among a set of BACs that are present within several colonies of bacteria that have been grown on a plate of nutrients. A colony results from one cell dividing over and over; therefore, each member of the colony will contain the same BAC. After a few hundred cells containing different BACs are spread on a plate of nutrient agar, the cells begin to divide and eventually create visible colonies on the plate. Each colony contains a different BAC. But because it would be very messy to attach radioactive probes to a colony of bacteria on the plate filled with agar, a semi-solid gelatin-like material, the scientists make a "print" or "lift" with a portion of each colony from the plate by putting a paper-like sheet over the colonies and lifting off an imprint of each colony (Figure 3.4). This creates a grid or pattern of the colonies on the sheet that mirrors the colonies on the plate. The probe will attach to a spot on the sheet that was in contact with a colony of bacteria that contains the same sequence. Therefore, the sheet is applied and lifted and then the STS probe is applied by soaking the sheet in a bag of liquid that contains the radioactive probe. The sheet is dried and then exposed to film. A dark spot on

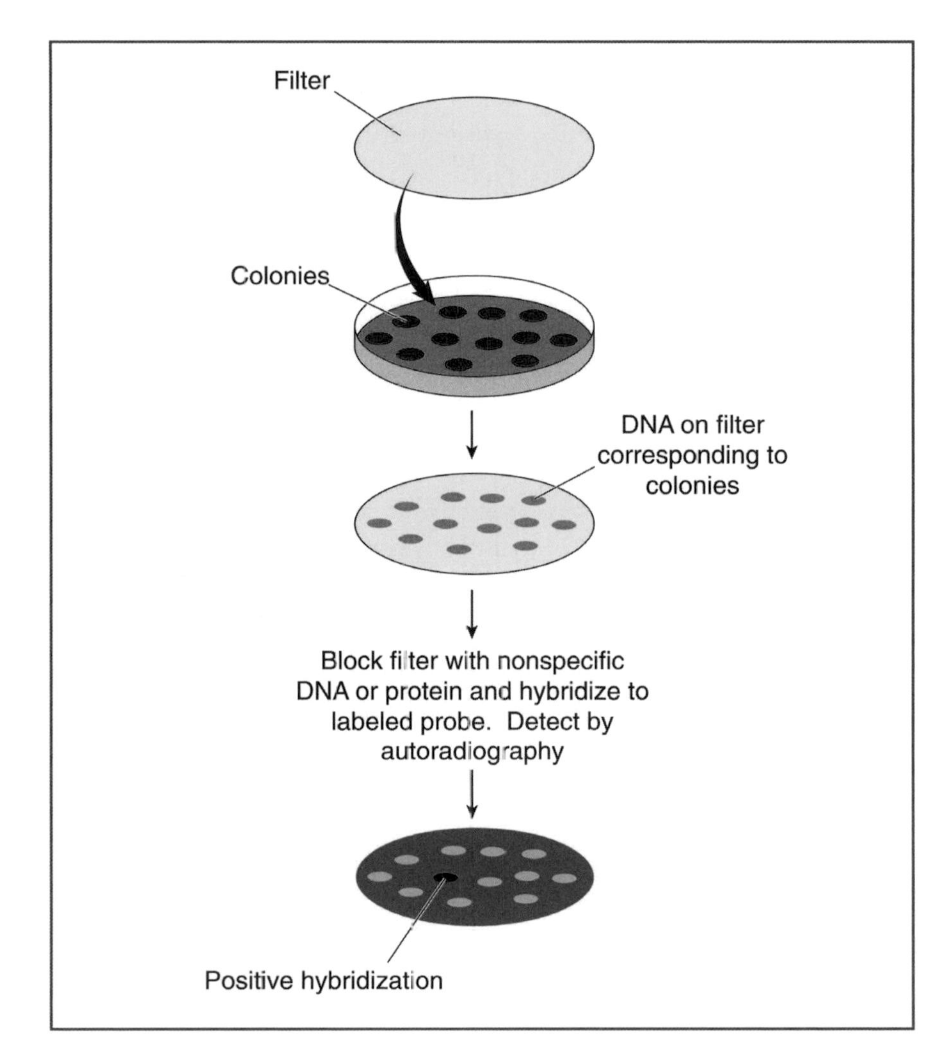

Figure 3.4 Colony hybridization (illustrated here) is used to identify bacterial colonies having a particular short sequence of DNA (such as an STS sequence) within a BAC. Bacteria containing different BACs are grown on a plate of nutrient agar, where they divide to form visible colonies of identical cells. The colonies are then "lifted" onto a piece of paper as a pattern of spots and a radioactive DNA "tag" is attached. The spot containing this DNA is made visible when the probe attaches by base pairing to its complementary strand of DNA and the paper is exposed to film.

the film shows where the radioactive probe attached. A match means that the STS sequence is present in the BACs of that colony. Because the STS sequence was present in that particular BAC, that BAC and all of the BACs associated with it in its contig must be positioned on the human chromosome around that STS sequence. In this way, the BAC and its contig were mapped to a particular location on a chromosome. Doing this over and over with different BACs and STSs the scientists, with the help of computers, could place many contigs on human chromosomes. Later, as the BAC DNA sequence was generated, they checked the positioning of the BAC and its contig along the chromosome by comparing the sequences of known STSs with the BAC sequence.

Fluorescent molecules emit light when a particular wavelength of light shines on them. Fluorescent DNA probes were made using BAC templates. Each probe contains some BAC sequence, and will therefore attach to its complementary DNA on a chromosomes spread out on a slide. This method can indicate on which chromosome there is a sequence similar to the BAC and tie the BAC to a particular chromosomal location.

Using these methods, Human Genome Project scientists ordered the BACs by clone-based physical mapping and then used STSs to tie them to particular regions of the human chromosomes. Then the BACS were shipped out to labs around the world to be sequenced. This strategy allowed labs to be sure they were not duplicating the efforts of another lab by sequencing the same section of DNA. Each lab then cut each BAC into several different batches of smaller pieces so there would be plenty of overlap and cloned these pieces into bacteria that would make more copies of them as the bacteria grew. They determined the sequence of the ends of the small pieces and arranged the overlapping fragments in order, largely using the **random shotgun sequencing** method described below, knowing that all the pieces were contained within a particular BAC. Ultimately the filled-in fragments from 20 labs around the world generated the public draft of the human genome, encompassing about 90% of each chromosome.

Remember, each of our 46 chromosomes contains just one long molecule of DNA. The Human Genome Project sequenced 24 of those molecules, one of each pair of autosomes or non-sex chromosomes and the X and Y sex chromosomes that determine whether a person will be male (XY) or female (XX). The two DNA molecules of each autosomal pair are almost the same. For each of us, one of the pair came from our mother and one came our father. Earlier studies of one gene at a time showed that these paired molecules have only about one difference in every thousand bases in the sequence. Therefore, scientists concluded they needed to sequence only one of each of these chromosomes. In addition, they sequenced one X and one Y chromosome, for a total of 24 chromosomes. In the end, the International Human Genome Sequencing Consortium (IHGSC) generated 23,000,000,000 bases of sequence. *Giga* means "billion" which comes from the Greek word for giant. Because scientists love to use abbreviations (perhaps you have noticed this?), they say that the Human Genome Project generated 23 gigabases or Gb of data or the equivalent of 7.5-fold coverage of the 3 Gb human genome.[6]

WHOLE SHOTGUN SEQUENCING

The private effort to sequence the human genome was led by Craig Venter and his company, Celera. At a 2001 genome conference, scientists bristled when the vice-president of Celera spoke. They were angry at his ability to use the public databases while competing with scientists in the public sector to generate a human sequence. Later it was learned that Dr. Venter's team had sequenced his DNA for the human genome sequence he generated, despite earlier protestations to the contrary, a move that exemplified his brash approach to this enormous effort. The random or **whole shotgun sequencing** process he used required enormous computer power (Figure 3.5).

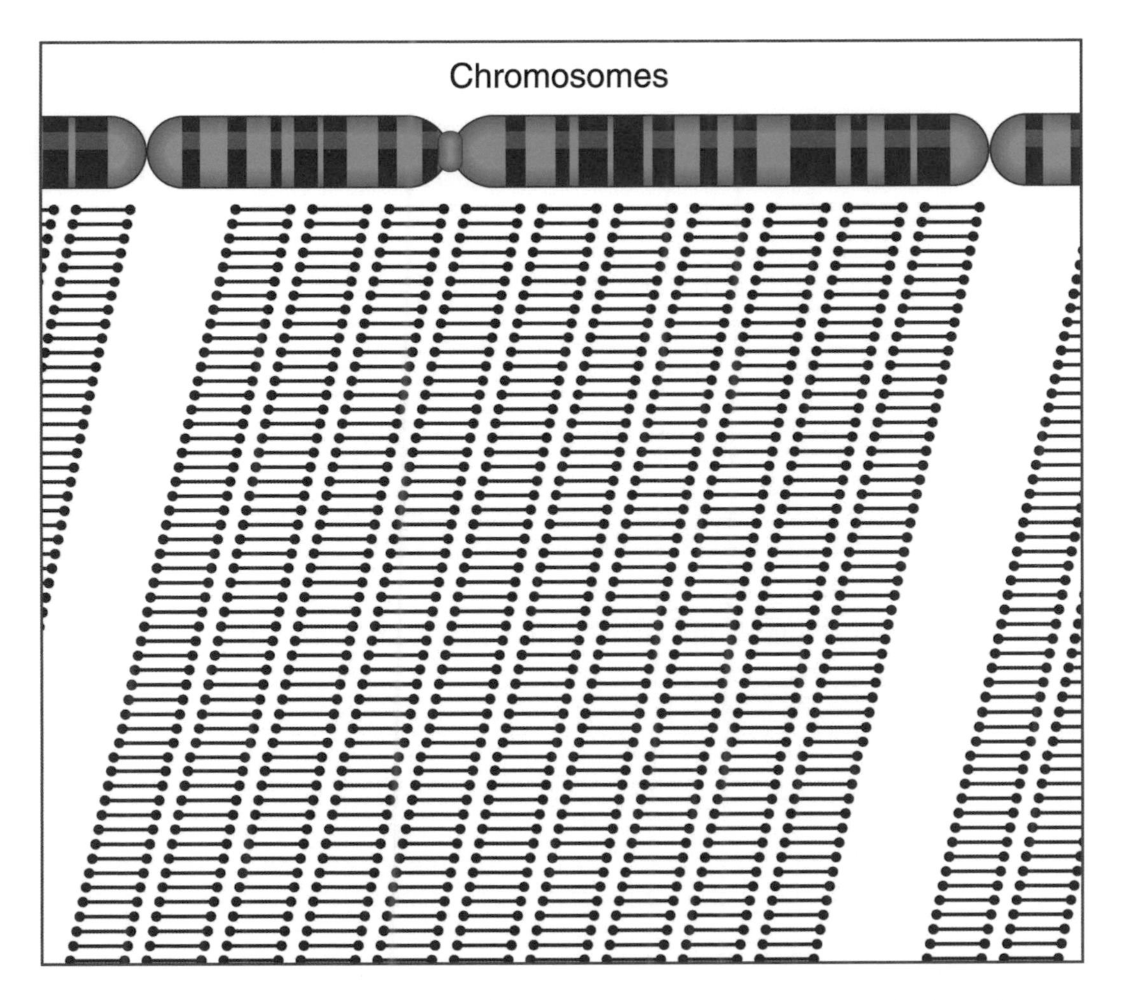

Figure 3.5 Craig Venter's group used the shotgun sequencing method illustrated here to determine the sequence of the human genome. They cut DNA into pieces of three different sizes and determined the sequences only on the ends. Then they used a computer to order the end sequences and in this way determine the sequences of each human chromosome.

Imagine having a computer put together a 300,000-piece puzzle where the pieces don't snap together, but rather the pictures on the pieces overlap. The puzzle depicts an English manor on a hill overlooking a stream with white clouds above. One can imagine that the hierarchical sequencer, the sequencer from the public genome effort, would first separate out the "sky," "manor," "hill," "stream," and "cloud" pieces. After separately matching together pieces to assemble each part of the puzzle, it would then link the sections together. With whole shotgun sequencing, there was no effort to separate out sections of the puzzle before assembling it. Imagine that most of the puzzle pieces had in fact earlier been cut into 3, 15, or even 75,000 pieces and mixed up. On all the pieces the picture appears only at the edge of the piece, but there are so many pieces that the edges overlap. For Celera's effort to sequence the human genome, several copies of Dr. Venter's genome were cut into 2,000 bp, 10,000 bp and 50,000 bp pieces and about 500 bp on each end was sequenced. Then the computer grouped all the overlapping pieces based on the end sequences to create sections or contigs of continuous sequence of the DNA puzzle from the random fragments. The large 50,000 bp pieces were not themselves sequenced. The final sequence mostly came from linking together the 500 bp end sequences of all the smaller pieces. The sequences on the ends, the edges, of the larger pieces served as a loose guide that the ends of the pieces in between were properly arranged. The end sequences tied to a particular piece of DNA were also used to link contigs together. The order of these contigs was known if one end sequence was in one contig and the other end sequence was in another contig. Therefore, the contigs had to be in a certain order, but the sequence of the gap in between was not known. In this way contigs of known continuous sequence were ordered into large blocks called scaffolds even though there were sequence gaps in between. As a final step, Venter's group tried to determine where on the chromosome these sequences actually occurred; in other words, they tried to match the scaffolds to the human chromosomes. This was done by consulting all of the STS sequences and other data linking sequences to particular chromosomal locations that had been published earlier. Whole shotgun sequencing was extraordinarily fast because the entire draft sequence of the human genome took Dr. Venter and Celera less than a year to assemble. It may not surprise you, after working through the logic of what they did, that most of their work was done by sequencing machines and by scientists using computers.

However, there was a good argument against using the whole shotgun sequencing method. Over 50% of the human sequence has repeat sequences that make mapping the genome difficult. Many of these repeats are from ancient hopping sequences called **transposons** and others are duplicated segments that may be next to the original sequence or may have been inserted into another chromosome. Imagine how almost identical pieces of the manor, hills, stream, and clouds would make it even more difficult to assemble the puzzle without noting first that some of these pieces are out of place because the piece may have ends that, for example, are part of a stream, but a middle section that is actually in a disconnected region such as the manor. This might make you pause. At the very least, examining such a puzzle piece would make you test more closely to see in which region the piece actually belongs. Without having such clues over a 150,000 base long stretch (the length of the public group's BACs) of what sequence is actually next to what sequence, the private group misaligned several repeats. One of the reasons that Venter's group succeeded was because they could use the sequences regularly deposited in public databases by the public team. Later analyses showed that, in fact, the Celera sequence missed many areas of duplicated sequence.[7] As the authors of the public sequence pointed out in their landmark paper, one of the main purposes of this effort is to accurately determine the sequence of gene mutations involved in disease. Therefore, even though the hierarchical sequencing method was slower, they felt it to be more accurate, especially in areas of repeated sequence. On the other hand, because large stretches of sequence have remained intact in us, the chimp, the mouse, and the rat over the past several million years since our most recent common ancestor, those genomes were largely completed by the whole shotgun approach using the human sequence and the sequences of other related organisms as a guide.

Stop and Consider

Compare and contrast hierarchical sequencing and whole shotgun sequencing. Under what circumstances might each be the better choice?

HIGH-THROUGHPUT SEQUENCING

Not too long ago, sequencing was laborious. The process required four separate radioactive reactions to indicate the positions of A's, T's, G's, and C's along one strand of DNA. The sequencing reaction is similar to a PCR reaction in that a DNA template, a primer (in this case a single radioactive primer), a DNA polymerase, and nucleoside triphosphates are needed for a new piece of DNA to be made opposite a single-stranded template. But in order to figure out the sequence, pieces that are one, two, three, four, five, and so on bases long must be generated and separated on a gel. How can this be done? In 1975, Frederick Sanger, an English biochemist who won two Nobel Prizes in chemistry, devised a method called chain termination. He added a small amount of dideoxynucleoside triphosphate to each reaction so that when added to a growing sequence would be the last nucleotide added (*dideoxy* means "without two oxygens". This is because DNA polymerization depends upon a free hydroxyl group (–OH) present on the sugar in each nucleotide. New nucleotides can be added to the growing chain only if they are added to the hydroxyl group.

Dr. Sanger decided to add a small amount of this unusual nucleotide to each reaction so that when incorporated into the growing piece of DNA, the dideoxynucleotide not containing this hydroxyl group would stop further growth in the chain. All four kinds of regular *deoxy*nucleoside triphosphates ("without one oxygen" but nonetheless having the critical one) were also present. Therefore, once the *dideoxy*nucleoside triphosphate was added to a sequence being made, the chain stopped growing. There were four reaction tubes. Each reaction had only one type of *dideoxy*nucleoside triphosphate but all four kinds of the regular *deoxy*nucleoside triphosphates. The G tube contained a little *dideoxy* G nucleoside triphosphate, the C tube contained a little *dideoxy* C nucleoside triphosphate, the A tube contained a little *dideoxy* A nucleoside triphosphate, and the T tube contained a little *dideoxy* T nucleoside triphosphate. These four reactions together generated all possible lengths of sequence but only the "G" tube contained pieces terminated at a G, only the "C" tube contained pieces terminated at a C, and so on. The most time-consuming part was preparing and then running the electrophoresis gel that separated the radioactive reaction products by size. Each tube produced a ladder of products in its lane on the gel. There was a "G" ladder, a "C" ladder, an "A" ladder, and a "T" ladder. The gel was then dried and exposed to x-ray film, the film developed, and the film was read by hand. Dark areas where the film had been developed from the radioactive fragments indicated the positions of each base in the sequence. The smallest pieces were the bands that had reached the bottom. The largest pieces were the bands near the top. To read the gel, you started at the bottom of the image. As you went up the image of the gel, you switched from lane to lane to see in which reaction lane the next larger band was found. In this way, it was possible to read the entire sequence of about 500 bases along one side of the DNA template that had been present in the reaction mix. This process was tedious! (Figure 3.6)

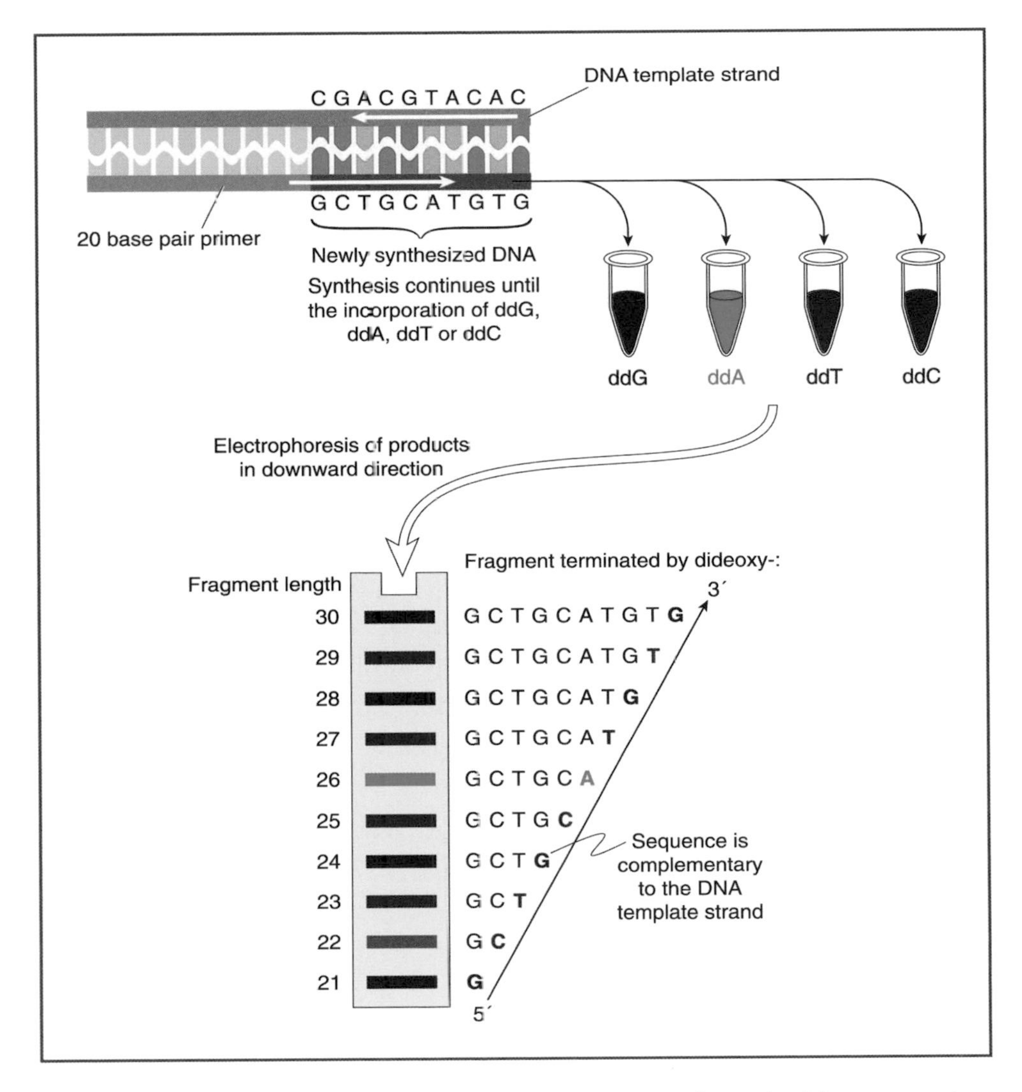

Figure 3.6 Previously, DNA sequencing was tedious and required four separate reactions as described in the text. Now a sequencing reaction requires only one tube. Each terminating nucleotide (ddNTP) has a fluorescent tag. The colored tag labels the end of each piece of DNA. The pieces of DNA generated using the unknown DNA as a template are separated on a gel. The fluorescent tags read automatically from the bottom of the gel provide the previously unknown DNA sequence.

By 1999, using high-throughput methods, every step in the process from generating clones to reading the sequence was automated. Without the automation and the speed and lower cost it allowed, the Human Genome Project would have been unacceptably expensive. The sequencing reaction itself had been simplified so that only one reaction, not four separate ones, was required. Four differently colored fluorescent tags labeled each reaction sequence and the reaction was amplified by PCR, the polymerase chain reaction. Ninety-six reactions were prepared at one time and loaded by robots onto a sequencing machine. As the fluorescent tags passed a detector, the sequence was read automatically by a computer. The basic reaction that generated the sequencing products was the same as before. However by introducing automation into every part of the process and taking advantage of the power of fluorescent labels and PCR, a single researcher running the ABI Prism™ 3700 automated sequencer could generate 500,000 bases of sequence in one day compared to only 500 bases per day using the old method.[8] This innovation was essential for completing the 3 billion base-pair human genome.

In February 2001, when Celera Corporation unveiled their draft of the human sequence in *Science*,[9] the draft generated by the public effort was presented simultaneously in *Nature*.[10] Although Dr. Venter had had access to all of the public data, he did not share Celera's data with his competitors. Nonetheless, his feat was remarkable. David Baltimore, himself a Nobel Prize winner as a co-discoverer of reverse transcriptase, the enzyme that allows the AIDS virus to make a DNA copy of itself and therefore infect our cells, commented in *Nature* on the draft sequence of the human genome. He said, "Celera's achievement of producing a draft sequence in only a year of data-gathering is a testament to what can be realized today with the new capillary sequencers, sufficient computing power and the faith of investors."[11] The sequence of each human chromosome determined by the cooperative effort by the public group in fact resulted in a more accurate draft of the human genome. Celera's whole shotgun method of sequencing bits without a roadmap mislaid several regions of duplicated sequence. But perhaps most importantly, the human genomic sequences obtained by the international consortium were available for scientists throughout the world to use.

CONNECTIONS

In 1866, Gregor Mendel reported his discovery of what we call genes to an unreceptive audience. Because this information lay dormant for decades, we have had only about 100 years to build on his findings. The push to determine the entire sequence of our chromosomes began toward the end of the 20th century and became a race only in its final years. The race was won at the dawn of the 21st century by both a public consortium of laboratories around the world and a private venture led by Craig Venter. Advances in high-throughput technology had allowed the private group to use enormous computing power and high-speed sequencing to assemble the entire human genome (admittedly missing some areas of highly repetitive sequences) by sequencing the ends of innumerable small pieces of DNA. Undoubtedly, the competition focused the efforts of the international consortium, which used a more deliberative approach. They used clone-based physical mapping to determine the overall assembly of BACs that covered the sequence of the human genome and then sequenced the BACS within those contigs by shotgun sequencing. Although the international consortium started earlier and therefore took longer to finish their draft of the human genome, in the end their version was more accurate. Most of all, the international group of laboratories were determined that the sequence of the human genome would not become the sole property of a company. Because of the efforts of the International Human Genome Sequencing Consortium, the sequence of the human genome is available for anyone to use, including you.

FOR MORE INFORMATION

For more information about the concepts discussed in this chapter, explore the following Websites:

Search the Web for "Nature genome gateway." Select this site and then select the link to the human genome. This site has been put together by the people at the scientific journal *Nature*. This Website provides links to many of the original research articles related to the human genome and also includes commentary that might be easier to read. In a section entitled "Meet the Authors" these Web hosts have posted pictures of people at the 20 centers around the world who were part of the International Human Genome Sequencing Consortium.

Type in the word "EcoRI" into your search engine to bring up illustrations of the action of *Eco*RI. The illustration at "Access Excellence of the National Health Museum" at *http://www.accessexcellence.org/AE/AEC/CC/action.html* clearly shows the function of this restriction endonuclease.

Use a search engine to look for information about "clone-based physical mapping." You should pull up the article by Maynard Olson, one of the main architects of the Human Genome Project from Washington University in St. Louis. This article is available free from *Nature* and is entitled, "The Maps: Clone by Clone by Clone," in volume 409 from February 21, 2001, pp. 816–818. Dr. Olson discusses clone-based physical mapping used by the public consortium. This article includes his original drawing of the process made back in 1981.

The process of gel electrophoresis is a fundamental tool of molecular biologists. Use your computer Internet search engine and type in "gel electrophoresis" to bring up a variety of descriptions of this process. A particularly helpful explanation of the process with clear pictures is found at the Molecular Biology Cyberlab at the University of Illinois at *http://www.life.uiuc.edu/molbio/index.html.*

In order to learn more about the use of DNA fingerprints in forensics, type the words "DNA fingerprint" into the search engine. There are many excellent Websites that explain the process by which a DNA fingerprint is made and used to determine guilt or innocence including Websites at Nova Online and at the DNA interactive Website at *http://www.dnai.org/index.htm*, via the links to "Applications," "Human Identification," and "Innocence."

Learn about PCR, the polymerase chain reaction for making many copies of the same section of DNA, at the Dolan DNA Learning Center of Cold Spring Harbor Laboratory at *http://www.dnai.org*. Select in turn the tabs "Manipulation," "Techniques," and then "Amplification."

Why Sequence So Many Different Genomes?

"I've seen a lot of exciting biology emerge over the past 40 years," Dr. David Baltimore of Cal Tech said in 2001. "But chills still ran down my spine when I first read the paper that describes the outline of our genome."[1] Dr. Baltimore was not easily impressed. He had shared the Nobel Prize in physiology or medicine in 1975 for the discovery of reverse transcriptase, the enzyme that copies the AIDS virus genome so it can live inside our cells. But he recognized the enormous leap forward that humans had made in determining the sequence of A's, T's, G's, and C's in 24 human chromosomes (our 22 autosomes plus the X and Y chromosomes that determine sex). What some have called the blueprint for a human being was laid out before him. Dr. Baltimore was commenting on the paper published in *Nature* by the public consortium of 20 laboratories throughout the world that had completed a first draft of the human genome; a competing version developed by a private corporation, Celera, and its head, Craig Venter, simultaneously appeared in *Science*.[2,3]

Only a little more than 1% of the 3 billion bases in the human genome actually codes for protein. Some refer to the rest as "junk DNA," but that is only because we do not yet know what it is doing. About half of our sequence consists of repeated sequences and most of those have been contributed by millions of ancient jumping elements called transposons, most of which thankfully stopped jumping 40 million years ago. Other repeats come from segments of chromosomes that have been duplicated during evolution and are either near their original source or now on other chromosomes. Our history is replete with such duplications, but many have been silenced by evolution as they accumulated mutations that made them unable to direct the synthesis of a protein. Other genomes like that of the bacterium *E. coli* are more compact. Almost all their sequence codes for mRNA and they have no introns.

We can use the metaphor of models of cars to depict the differences among the genomes of living things. Twenty-four Hummers™ represent the 24 kinds of human chromosomes (22 pairs plus an X and Y) in which we drive to the mall in a display of excess, given that so little sequence actually codes for proteins. By analogy, the four fruit fly and six *C. elegans*, or worm, chromosomes are a fleet of SUVs. They also contain "junk DNA" but less of it than ours, and only about 5% of their sequence are transposon repeats. The 16 yeast chromosomes are the Camrys™ of the fleet. They have only one gene for every 2,000 bases, almost no introns, and fewer than 100 transposons. The *E. coli* bacterial genome is the Volkswagen® bug on the road. It has about 1,000 bases per gene lined up one after the other. Its genes have no introns. To save space and coordinate expression, some *E. coli* genes even share the same regulatory regions to create **operons**. Viruses have only a handful of genes and some of them overlap! (Imagine what is required of the DNA sequence to make that happen!) Unable to reproduce on their own, viruses need to hitch a ride in one of the cars in order to go to the mall.

WHY SEQUENCE GENOMES?

The drive to sequence not only the human genome but those of a variety of other model genetic organisms is not purely utilitarian. It is true that knowing the sequence of the human genome will allow us to more accurately pinpoint the molecular cause of inherited disorders and disease. It is also true that knowing the genomes of other organisms such as flies, worms, and mice will allow us to manipulate genes that are homologous (similar) to the human disease genes in a living organism whose physical and chemical environment we can control. But another compelling reason for learning the DNA sequence of a variety of genomes intrigues researchers, and that is to take a fresh look at the tree of life. Anytime scientists compare one genome with another, they are making an assumption that the two genomes share a common ancestor and that the differences between the two genomes reflect changes that have occurred over time. At each position within the chromosome where a base substitution occurs, for example an “A” in one genome where there is a “G” in another, one can make an assumption about which was the original sequence. The genome of a closely related organism assumed to be derived from older stock can help to make this choice more obvious. By literally counting the changes that must have occurred along the tree’s branches, one can construct a molecular tree of life. Remarkably, these trees parallel to a large extent those that were based on the fossil record, but they allow us to make distinctions for which no other record exists (see Figure 4.1 for a timeline of common ancestors).

WHAT DID THE HUMAN GENOME REVEAL?

By the time the draft of the human genome was unveiled, scientists already knew the complete genomes for several viruses, bacteria,

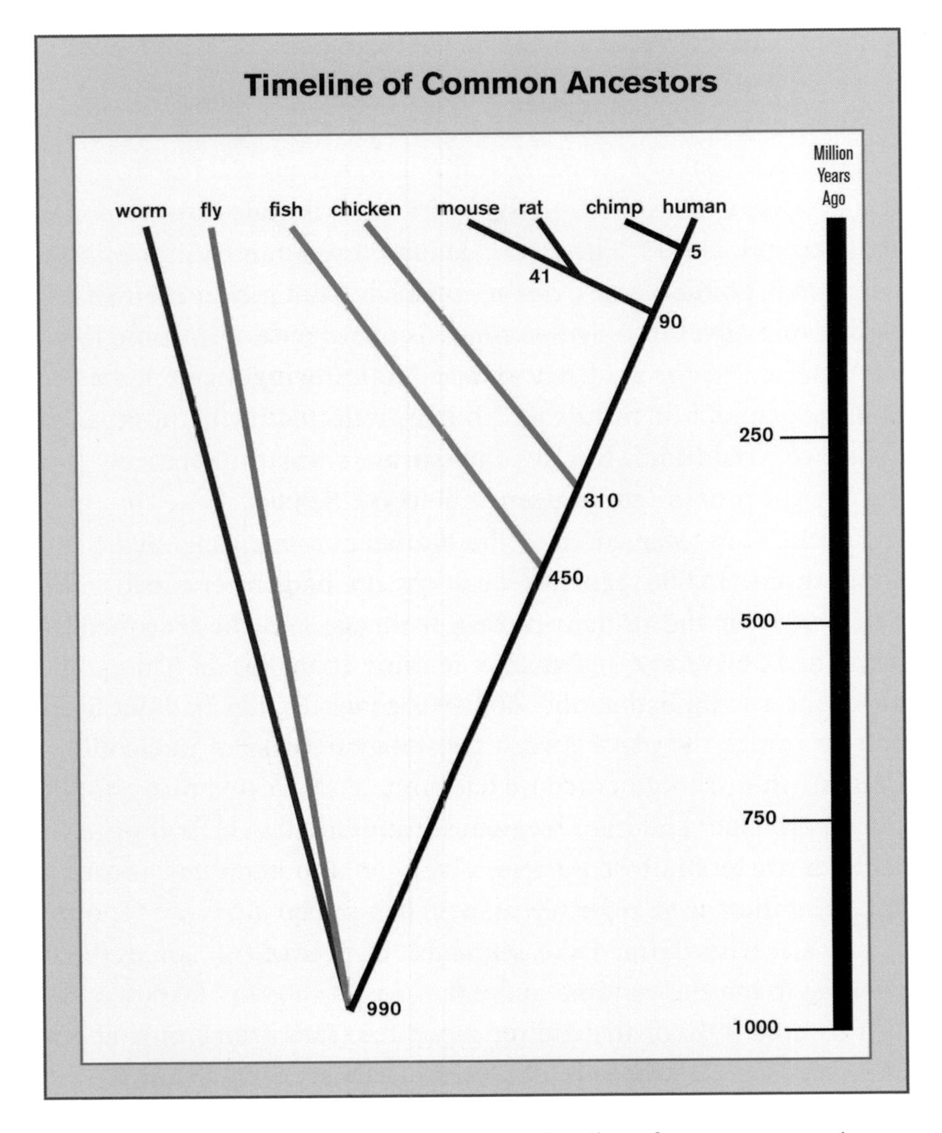

Figure 4.1 Shown here is an example of a timeline of common ancestors. Based upon both DNA sequences and the fossil record, the common ancestor of many of today's animals lived about a billion years ago. This chart can be used to see when certain types of organisms arose. (The data are largely derived from S. Blair Hedges, "The Origin and Evolution of Model Organisms," *Nature Reviews/Genetics* 3 (2002): 838–849.)

yeast, and invertebrates, organisms that spanned over a billion years of evolution. Viruses that rely on a host for most of their needs encode only a handful of genes. The bacterium *E. coli* has about 4,000 genes, baker's yeast has about 6,000 genes, the fruit fly 14,000, the worm *C. elegans* 19,000, and the weed *Arabidopsis thaliana* has 21,000. The *E. coli* genome is contained in 5 million bases. At 3.2 billion bases, our genome has 640 times as many bases as that of *E. coli* but only 5–6 times as many genes. On some level, we expected to see evidence of our "superiority" or at least the complexity of our minds and bodies reflected in the number of genes we would find. But our first surprise was the relatively low number of protein-coding genes, 20,000–25,000.

Both "draft" sequences of the human genome suggested there were about 30,000 genes.[4,5] Each group had determined only about 90% of the protein-coding sequence and these sequences contained on average one mistake in every 10,000 bases. The public group's draft still had about 150,000 gaps where they had not been able to bridge the gap between two long stretches of nucleotides. Celera's whole shotgun method had completely missed many duplicated segments[6] and many segments were misaligned or flipped in both of these draft sequences.[7] Despite the room for error, it appeared there were no more than 30,000 genes.

As scientists refined the sequence, their and our amazement grew. By 2004, the genome had only one mistake in 100,000 bases and covered 99% of the coding sequences. There were only about 300 gaps left in the public sequence. This clarification of the original sequence shows there are only 20,000–25,000 genes in 2.9 billion bases of sequence.[8] (We have still not been able to decipher sequences near the **centromeres** where there are many repeats and almost no genes. Centromeres are the regions on each chromosome that attach to the cellular strings that tug and make sure that when a cell divides, each daughter cell gets one copy of each chromosome. Resolution of the sequence near the centromeres must await advances that allow these sequences either to be stably cloned or read by a completely different means.)

In some ways it is shocking that we have so few genes. This number, 20,000–25,000, is only about four times the number found in yeast and is comparable to the number found in the roundworm, puffer fish, chickens, rats, mice, chimps (Figure 4.2), and even in the weed *Arabidopsis thaliana.* Past predictions had ranged from 20,000 to over 100,000 human genes, but the prediction that had taken hold was a "back of the envelope" rough estimate of 100,000. Scientists often make such rough estimates, based upon the best evidence on hand, but they often do not expect them to be repeated. However, this figure made its way into the collective consciousness largely because it was often quoted in textbooks. How was that estimate made? Based upon the sequences of human genes that had been decoded one at a time, the average gene was known to be about 30,000 bases long. Since there were about 3,000,000,000 bases in the entire genome, W. Gilbert[9] had made the offhand estimate that this would leave room for no more than 100,000 genes (3,000,000,000/30,000 = 100,000). So, an estimate proposed as an upper limit on gene density became the number of genes in the human genome most often quoted. This estimate did not take into account the three-quarters of the genome sometimes called "junk DNA" that does not code for genes.

The draft sequences of the human genome have also allowed us to more precisely determine other statistics about our genes. On average each of our genes has about 9 exons, which is similar to other vertebrates, with 8 exons in the average mouse gene and 7 in the average fresh water puffer fish gene. In contrast, most yeast (*S. cerevisiae*) genes have only one exon. For bacteria, the question is moot: each gene is coded for by a long string of bases, with no interruptions. While each of our exons is actually quite short (on average about 150 bases long), we differ from invertebrates most in

ORGANISM	HAPLOID* GENOME IN BASE PAIRS	NUMBER OF GENES
E. coli (bacterium)	4,700,000	4,300
S. cerevisiae (baker's yeast)	12,100,000	6,200
C. elegans (worm)	97,000,000	19,000
D. melanogaster (fruit fly)	180,000,000 (only 120,000,000 sequenced)	14,000
T. nigroviridis (freshwater puffer fish)	340,000,000	22,000
G. gallus (chicken)	1,000,000,000	20,000–23,000
M. muscularis (mouse)	2,500,000,000	<30,000
R. norvegicus (rat)	2,800,000,000	21,000
P. troglodytes (chimpanzee)	3,100,000,000	20,000-25,000
H. sapiens (human)	3,200,000,000	20,000-25,000

* Haploid cells have only one set of chromosomes (unpaired).

Figure 4.2 The genomes for over 1,000 organisms have been sequenced. This means that the DNA sequence of one of each homologous pair of chromosomes plus the sex chromosomes have been determined. Above is a list of the genome sizes (in base pairs) and the number of predicted genes for a variety of organisms.

the length of our introns, which are on average over 3,000 bases long. The average fly and worm introns are only 500 and 300 bases long respectively. This makes our typical gene quite long, close to 30,000 bases. However, our longest gene, which codes for the protein dystrophin, contains 178 exons! Its longest exon is about 17,000 bases. The entire gene spans an amazing 2,400,000 nucleotides. When the dystrophin gene is mutated, so that no dystrophin is made, a person can develop a condition called Duchenne muscular dystrophy, which leads to weak muscles.

Researchers have determined that we have more proteins than genes. How can this be so? One way we generate protein **complexity** is by having one gene code for several different mRNAs by cutting out different introns (and exons with them). According to a relatively small survey along only one chromosome, we make 2.6 different transcripts for each gene, compared to 1.3 made by the worm. Perhaps most importantly, given our lack of superiority in the number of genes, we trump our evolutionary relatives in both the invention of new **protein domains** and the way that we have reused old domains. Domains are the structural and functional units of proteins in three-dimensional space. In Chapter 7, we will explore the domains of a key protein shared by many different cells and many different organisms.

Gene Family Expansion

Because protein domains provide the structural basis for protein function, it is useful to think about domains as units that have been targets of natural selection and evolution. It appears that domain invention and shuffling has come about through a variety of means. A particularly successful process has been duplication of small and large segments of chromosomes. These duplications have allowed proliferation of a family of genes, where one gene can retain the original function and other copies are allowed to evolve on their own time until random changes generate the plans for a useful protein that has many of the same functions but with adjustments in sensitivity or its ability to interact with other cellular machines. It is striking that many large and small portions of our chromosomes have been duplicated and reinserted either into the same chromosome or into different ones. Many of these duplications are thought to be the work of transposons that copied portions of adjacent genes by mistake and moved those domains with them when they hopped, or that mistakenly cut and paste, large segments of chromosomes, not just their own DNA. In fact, an analysis of the draft genome sequence shows that contrary to the idea that dropping a transposon sequence into a coding region is the kiss of death, many known and predicted proteins contain at least partial sequences from transposons.[10]

Both human genome research groups noted that various families of genes expanded in humans relative to the proportion of similar genes in invertebrates and those in unicellular yeast and bacteria. Two of the most remarkable comparisons presented in Dr. Venter's *Science* paper are contained in two ordered lists. One is a list of specific **structural domains** and profiles with functional implications found in the human, fly, worm, yeast, and weed genomes, organized by broad categories of cellular function. These are called Pfams. The second is a list of proteins of related function present in these completed genomes. For example, proteins peculiar to animals regulate animal embryonic development and programmed cell death, a process important in normal development. These proteins have either no or very few representatives in yeast and in plants. In contrast, there is a tremendous expansion in genes coding for proteins that have these functions in the human genome. Similarly, although there are multiple genes programming only one or two kinds of structural domains for the immune response in plants, flies, and worms, there are multiple kinds of domains and many genes encoding them for the immune response in humans. Since the human genome has only four times as many genes as yeast, the mere increase in the number of genes cannot explain this explosive proliferation.[11]

The public genome effort commented on the expansion of domain structures demanded by the predicted environment of those proteins. The unicellular yeast's genome, which was solved in 1996, is full of sequences for proteins that have **intracellular** (intra = within) domains with only a few domains crossing the membrane (**transmembrane**) or extending beyond the cell (**extracellular**). Given that yeast's survival depends upon its abilities to find food and use this food to generate all of its internal structures, having many genes devoted to enzymes that tear apart and put together food molecules is appropriate. Nonetheless, both the fly and the worm have increased their stock of intracellular domains over those found in yeast. In addition, these multicellular invertebrates have a greater variety of protein domains that span their membranes and extend outside of the cell. (Can you imagine why this would be advantageous in a multicellular organism?) But the variety of "distinct domain architectures" present in humans is greater in all of these spaces, allowing us to perceive an incredible variety of chemical, electrical, pressure, and light stimuli on the surface of cells and expand exponentially the possibilities for cell-to-cell communication. Dr. Eric Lander and his coauthors of the human genome produced by the public consortium also commented on the increase in the number of the members of many families of proteins in humans and other vertebrates over those present in invertebrates.[12] For example, humans have 30 different kinds of growth factor genes that stimulate certain cells called fibroblasts to grow, while the fly and the worm have only two each.

You may be surprised to learn that based upon both the fossil record and the incremental changes in the sequences of certain genes common to all forms of life, humans and chickens shared a common ancestor 310 million years ago. We shared a common ancestor with a typical fish like the puffer fish over 450 million years back. Even though the fish has about one-tenth the amount of DNA that we do and the chicken has only one-third of our DNA, we all have about the same number of genes, 20,000–25,000. This decrease in the chicken DNA nucleotide number is reflected in fewer transposons, duplicated segments, and **pseudogenes.** Pseudogenes appear to be mRNAs that used a reverse transcriptase to copy themselves into DNA and inserted themselves into a chromosome. About 60% of chicken genes can be found in humans and on average they are only 75% similar to their human counterparts. About three-fourths of these genes shared between chickens and humans are also found in fish.[13]

In addition, there are long blocks of conserved segments of synteny or regions where the same genes are aligned along parts of chicken and human chromosomes[14] and along fish and human chromosomes.[15] Figure 4.3A shows how probes corresponding to human chromosomal sequence flag their counterparts along dog chromosomes and Figure 4.3B indicates conserved segments of synteny between cat and human genes. While we share similar genes with regard to basic cellular structures and functions, chicken and fish genes that pertain to reproduction, immunity, and particular responses to their environment are, not unexpectedly, different from ours. For example, chickens have more odor receptor genes but fewer taste receptor genes than we do. We lack their genes for making egg whites and they lack ours for producing milk proteins. Medical researchers are intrigued that the chicken has a gene that codes for interleukin-26 (IL-26), an immune-response protein previously known only in humans but which investigators may now study in the chicken. The corresponding genes in both mice and rats have inconveniently been inactivated by mutations.[16] Not surprisingly, fish have a greater

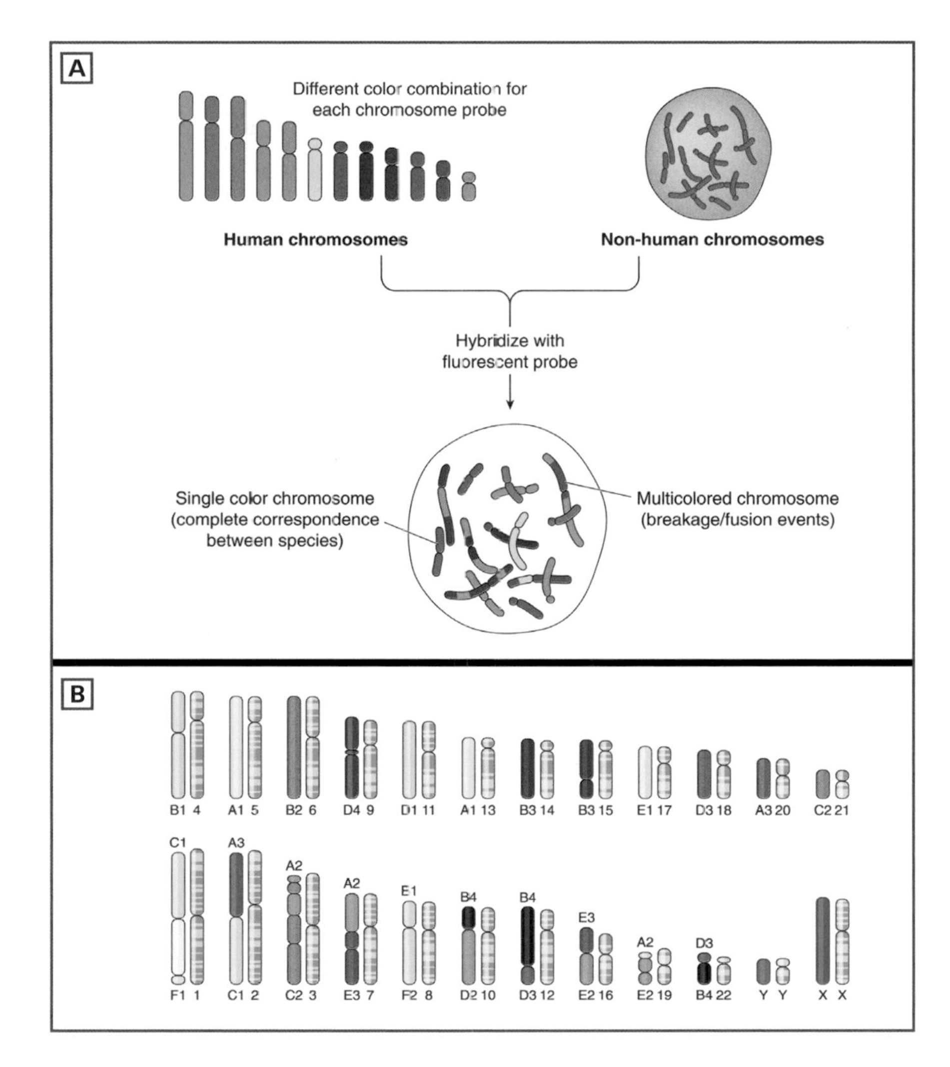

Figure 4.3 Shown here is an example of chromosome painting and conserved regions of synteny. (A) Colored probes that correspond to human chromosomes use DNA pairing rules to indicate regions within non-human chromosomes that are similar. (B) Thirty-eight cat chromosomes were combined into six chromosomal groups, each a different color. One can see how the sequence order has largely been preserved in at least 14 human chromosomes (shown in blue) and the corresponding cat chromosomes, including the X and Y, since our last common ancestor. *Source for (B): W.J. Murphy, et al., "A Radiation Hybrid Map of the Cat Genome: Implications for Comparative Mapping,"* Genome Research *10 (2000): 691–702.*

number of sodium pumps for regulating the salt content of their cells than do humans.[17]

Human, Rat, Mouse Comparisons

A three-way comparison of the human genome with those of the mouse and rat, among our closest mammalian neighbors, reveals striking similarities. Evidently, we long ago intuited our kinship, because we have used rats for almost 200 years to learn about cardiovascular disease, various psychiatric disorders, cancer, diabetes, and so on, and often test the toxicity of drugs in rats before testing them in ourselves. Through the years, we have developed over 230 inbred strains of *Rattus norvegicus* with particular forms of disease genes so that the rat can serve as a model for an inherited human disease. Similarly, we have developed mouse models to study human disease and have created numerous "knockout" mice that allow us to study the effect of these genetic changes on disease. *Knockout* is the shorthand term for an organism in which a gene has been inactivated or removed. Comparisons of the genomes show that we share nine out of every ten of our protein-coding genes with *both* rats and mice and that each gene we share is about 90% similar on average.[18] As is true of our genes and those of a more distant relative, the cat, not only do we share the same genes, but our genes are largely co-linear in chunks of sequence along the chromosomes. These pieces may be in the same orientation or they may be flipped. In some cases, the shared genes are spread out among several of our counterpart's chromosomes. Before the mouse and human genomes were sequenced, we used this knowledge that many genes were grouped together and in a similar order to help us locate human genes, once we had identified the corresponding mouse gene. Comparison of the mouse, rat, and human genes shows that about 3% of the rat genome arose from segmental duplications, an intermediate amount relative to those in mice (1%–2%) and humans (5%–6%).[19] Not unexpectedly, these duplications have given rise to new genes so that some of the rat's genes for detecting pheromones (chemical sexual attractants) and enhancing immunity are different from those in both the mouse and humans. The detoxification genes are of special interest because both rats and mice are used as model organisms in drug safety studies. One of these genes called CYP2J has only one copy in humans, but four in mice and eight in rats, which may account for differences in susceptibility of mice, rats, and humans to certain drugs.[20] Sometimes animals used in the laboratory are more resistant to certain drugs because their bodies can get rid of the drugs more easily.

Stop and Consider

Why are conserved domains across species important for genetic research? What do these conserved domains tell scientists?

Human, Chimp Comparison

We humans continually seek to understand what makes us human, and comparative genomics can begin to answer that question, especially with regard to how we are *different* from the chimpanzee.[21] Before embarking on a discussion of the distinctions between chimps and humans, it is important to realize that the DNA sequences of our genes differ in only one out of every 100 nucleotides with those of a chimp. By comparison, your DNA sequences differ from any other person (except a family member) on average by only one out of every 1,000 nucleotides. Another 5% of alignment differences between chimps and humans come from insertions and deletions since the chimp and human sequences diverged.[22]

A comparative study of chimp, human, and mouse genes was conducted in December of 2003 by a team of Celera researchers and scientists from Cornell University.[23] Such studies examine or reveal genes that code for key differences between species. The results indicated that genes for smelling and hearing have undergone changes that have been crystallized into permanent coding differences between chimps and humans. For example, a human gene whose difference from the chimp gene has been carefully maintained by evolution over the last 100,000 years codes for a protein found in the inner ear called α-tectorin. People who have a particular amino acid change that makes their α-tectorin protein more like the chimp's protein have trouble hearing high-pitched sounds.[24] Another difference between chimps and humans is the code for a "speaking gene" called *FOXP2*. A few changes in the base sequence of this gene occurred sometime within the last 200,000 years and then the change quickly became the norm in humans. This gene was recently identified by a mutation in people who find it difficult not only to speak, but also to naturally incorporate rules of grammar into their speech! In fact, when this gene was sequenced in people around the world whose speech was normal, it was found that everyone tested had the typical human sequence and not the chimp sequence.[25] It seems reasonable that genes that allow us to speak and hear sounds typical of speech may be some of the differences that help to make us human.

PRO OR CON?

Two known differences between chimps and humans are in the α-tectorin protein that affects our hearing and in the *FOXP2* gene that affects our ability to speak and use language properly. One way to see the effect of the human genes in a chimpanzee would be to try to change a chimp's genes into the human forms. Would it be ethical to try to make a chimpanzee more like a human being in this way? Why or why not? Do you think that the chimpanzee would acquire the ability to speak and use language properly? Why or why not?

Our ancestors had about 1,000 working olfactory receptors (protein smell detectors) that have remained active in species like the mouse and dog. But many no longer work in both chimps and humans. Nonetheless, some of our olfactory receptors have continued to undergo positive selection relative to the chimp genome; in other words, chance human mutations have become the norm in humans, possibly reflecting the odors that we find attractive that chimps do not.[26] There are other ways in which apes and humans differ. For example, the great apes only rarely progress to AIDS after infection with HIV and are resistant to malaria caused by *Plasmodium falciparum.* Apes rarely go through menopause or succumb to Alzheimer's disease. Some scientists would like to mine the human and ape genomes for differences that can explain these effects.[27] However, separating the signal (the important sequence differences) from the noise (the irrelevant differences in the 1% of amino acid differences between our proteins) will take some ingenious strategizing that can narrow down the choices.

Earlier comparisons of the chimp and human genomes had suggested that the major differences lay in the changes in the regulatory DNA sequences surrounding the genes and not in the genes themselves. Regulatory DNA sequences are places where proteins attach to the DNA. These interactions between cell protein and the regulatory sequences determine under what circumstances, at what time, and in which tissues mRNAs are made. It would be simple if the off and on switches for genes were the core of the differences, but a more complicated picture is emerging that suggests that other proteins, such as the proteins involved in hearing and speaking, and not merely the sequences and proteins that regulate when and where mRNAs are made, allow you—but not a chimp—to read this chapter.

CONNECTIONS

We have determined the base sequence of an assortment of genomes that span over a billion years of evolutionary history. These sequences will provide not only insights into what makes all of us different, but also in what ways we are the same. Already we have strengthened earlier insights concerning our origins and have learned more about the structures and evolution of our genomes. While our initial goal was to understand inherited disease and inherited disorders, we have a newfound appreciation of our chimpanzee cousins and long-time lab partners, the rat and mouse. Comparative genomics may allow us to mine our differences with chimps both to enhance our resistance to disease and help us to better understand what makes us human.

FOR MORE INFORMATION

For more information about the concepts discussed in this chapter, explore the following Websites:

If you want to explore more about different genomes, you can access worm and fly genomic information at a number of Websites. Although such information is available at *www.ensembl.org* and at the NCBI Website *http://www.ncbi.nlm.nih.gov/mapview/*, there are special browsers called Flybase (*http://flybase.net/*) and Wormbase (*http://www.wormbase.org/*) that allow one to search for additional specific fruit fly and *C. elegans* information.

Type in "Pfam" and use your search engine to search for the Pfam database at Washington University in St. Louis. Select "Help" and the "Frequently Asked Questions" to learn more about Pfam protein families.

You can access human and chimp genomic information at *www.ensembl.org*. To check out the *FOXP2* differences, type in *FOXP2* next to the "Lookup" button. Make sure the selection is "All" next to "Find" and select the "Lookup" button. Use this Website to explore the *FOXP2* gene and the

715 amino acid human protein. Note that this protein has fixed two amino acid changes in exon 7 at threonine-to-asparagines and asparagine-to-serine changes at amino acids 303 and 325 in comparison with the chimp (*Pan troglodytes*) protein.

5

Medicine in an Age of Proteomics

"It's akin to Eve taking a bite of the apple. Once you have the knowledge, there's no turning back," wrote Jessica Queller, a writer for the TV show *Gilmore Girls.*[1] Ms. Queller was not talking about her creation, Rory Gilmore, who had recently renewed her relationship with her married former boyfriend, Dean. Rather Ms. Queller's statement was about the difficult choices facing her after breast cancer took her mother's life. Ms. Queller had chosen to be tested for the flawed BRCA1 gene that had substantially increased her mother's risk of breast cancer; her younger sister had chosen not to be tested. Ms. Queller, who now knows that she carries her mother's BRCA1 mutation, must consider options for treatment (her BRCA1 mutation is also associated with an increased risk of ovarian cancer) to enhance her chances of survival. She is 35, still dating, and wants children.

Much of the public support for genome studies has been fueled by the desire of the public and medical researchers to understand and counter genetic sources of disease. Scientists pursuing basic research, on the other hand, are usually driven by a curiosity to understand life in all its forms. Increasingly, molecular biologists want to understand the networks of interacting molecules that make up life. Out of these complimentary goals, scientists are beginning to develop the tools not only to create a gene index that tallies the effects of any variation in the DNA sequence for every gene in man and other organisms like the chimp and mouse, but a protein index for every protein made by man and other organisms. But the goals of the Human Genome Project (HGP) are much more complicated than that. On a first level, the scientists and sponsors of the HGP want to understand the extent of variation in human genes and the proteins coded for by the genes. In addition, they want to understand in which tissues and under what conditions these proteins are made. Scientists have begun to appreciate that a snapshot of the protein profile in an organism or even a cell doesn't account for the fact that each cell and organism is a dynamic system of interacting proteins and the products of those proteins. So, more ambitiously, the HGP researchers would like to understand the molecular networks of proteins and protein products that operate under both "normal" conditions and in a diseased state. Finally, they want to provide the foundation for the development of drugs that target a particular molecule within an interacting network to correct an inherited disorder, whether it is a disease caused by one gene and its defective gene-product such as cystic fibrosis or a disorder such as certain cancers or cardiovascular disease that can result from the interactions of multiple gene products and a lifetime of environmental effects.

ONE TYPE OF VARIATION IN DNA: SNPs

One place to start is to gain an understanding of how we humans differ from one another. Merely creating a gene database that catalogues the variation present in the DNA sequences of human beings that populate the world is an enormous task. In the Howard Hughes Holiday Lecture of 2002 called "Scanning Life's Matrix: Genes, Proteins and Small Molecules" (that incidentally you can download or order free at *www.dnai.org*), Professor Eric Lander discusses the generation of such a catalogue.[2] The first goal is to determine the extent of variation within the human genome. This variation is largely (but not entirely) evident in the single nucleotide substitutions that make your DNA molecules different from those of your neighbors by having, for example, an "A" at that position instead of a "C." Because there is on average one nucleotide difference in every thousand bases, you will have on average 3,000,000,000/1,000 x 2 (because you have two of each type of chromosome) = 6 million differences from either the person sitting next to you or a student in Kyoto, Japan. (It doesn't matter who the other person is: the extent of variation is pretty much the same throughout the world.) Each of these differences is called a **SNP** (pronounced "snip") or **single nucleotide polymorphism.** One can see that if a SNP is in a gene's coding region, there is the potential to change which amino acid will be inserted at that position in the code for that protein.

Ms. Queller, mentioned at the beginning of this chapter, learned that she carried her mother's faulty BRCA1 gene by taking a test. This test considers a range of possible changes that includes not only SNPs but also deletions of a part of the gene. Details on how these tests are done and background on the search for the BRCA1 gene can be found at www.dnai.org.

The BRCA1 gene encodes a protein that has several roles within the cell, roles that make it a **tumor suppressor.** Intact tumor suppressor genes normally function to put the brakes on the growth of cells. One of the ways it does this is to help to repair broken DNA in anticipation of cell division. DNA can become damaged by the effects of x-rays, UV light, and some chemicals. To trigger the repair, BRCA1 must attach to an enzyme called DNA helicase through its C-terminal BRCT domain. Recently, three crystal structures were determined that highlight the interaction between BRCA1's two BRCT domains and the DNA helicase.[3,4,5] (You can find information about the BRCA1 gene using the NCBI Website, which is explained in Chapters 6 and 7, and also in the Appendix.[6]) Geneticists and biochemists have recently developed an idea of how some BRCA1 mutations put the cells at risk of becoming cancerous. The C-terminal domains of the BRCA1 protein, the BRCT domains, cradle proteins like the DNA helicase in a cleft between them. A signature sequence of four amino acids of a protein like the DNA helicase allows it to snuggle up to one of the BRCT domains as if it were a soft pillow. Several of the conserved amino acids in the intersection of the two BRCT domains that contact the helicase are mutated in cancer-causing alleles of the BRCA1 gene. For example, the positively charged amino acid arginine 1699 in BRCA1, when changed to a tryptophan (R1699W), is linked to cancer because with the hydrophobic tryptophan, the BRCA1 no longer provides a tight fit with the DNA helicase. Because of the failure to bind to DNA helicase and other proteins that are critical for its role as a tumor suppressor, damaged DNA will not be repaired by the mutant BRCA1 (and perhaps the brakes before mitosis will be released prematurely allowing broken DNA to be distributed incorrectly) and the person carrying this mutation will be at risk for breast cancer.

SNPs can also be located in parts of the gene that do not code directly for amino acids in the protein. These regions of the gene, called regulatory sequences, are often located either within or surrounding the DNA that codes for a particular mRNA. Regulatory sequences help to determine how much mRNA will be made and in which tissues. They also determine which parts of the mRNA will be spliced out and therefore which domains will be incorporated into the protein made. A SNP that creates a splice site where none existed before would make the DNA code say "cut me out" instead of "keep me." For example, the disease that causes small children to appear aged and develop the diseases of elderly people before their time, Hutchison-Gilford Progeria (derived from the Greek word *geras* for "old age"), has recently been linked in two different studies to a single change in the code for a protein called Lamin A that changes a C base to a T base.[7,8] Lamin A is a protein in the membrane of the cell's nucleus. This SNP in fact does not change the glycine that is normally coded for at that position. The code for amino acid number 608 is still a code for glycine. But this change in exon 11 makes the splicing apparatus, at least sometimes, cut out the last half of this exon and eliminate 50 amino acids from the end of the Lamin A protein.

Since this protein, normally found in **nuclear membranes**, helps to bring other proteins to the nuclear membrane, the nuclear membranes of children who have this disease are malformed and fragile, and may allow some of the DNA to leak out of the nucleus. Evidently cells that must put up with a lot of stress such as those in the muscles and skeleton can no longer function with this weakened nuclear membrane. The children with this mutation are small and usually die by age 13 from coronary artery disease or stroke. Because these children die before having children of their own, this SNP is not a normal variant present in the human population. The researchers compared the genomic sequence for this protein in the children with the sequence in their parents in order to see if this particular change had been inherited. It had not been inherited, but rather the sequence change had arisen as a spontaneous mutation. Dr. Francis X. Collins, who is the director of the National Human Genome Research Institute and the leader of one of the competing research teams who found the defect in the Lamin A gene, noted:

> Finding the gene for Progeria would have been impossible without the tools provided by the Human Genome Project. This was a particularly challenging project of the gene hunters, since there are no families in whom the disease has recurred, and geneticists generally depend on such families to track the responsible gene. This was a detective story with very few clues.[9]

Dr. Eric Lander, who was the first author of the published report of the public human genome sequence, would like to catalogue SNPs so as to correlate DNA sequence differences with the appearance of particular diseases. In the Howard Hughes Medical Institute Holiday Lecture of 2002, Dr. Lander discusses his dream of generating a chart that has our 22 autosomes and the X and Y chromosomes across one axis and diseases along the other (Figure 5.1). This chart would indicate all of the SNPs in each chromosome that are uniquely associated with each particular disease. In fact, physicians are beginning to correlate a constellation of such changes with particular disorders. In this way, they hope to understand a variety of disorders that are the result of not just one but several different genetic changes. For example, it has been notoriously difficult to pinpoint the constellation of genetic factors that predispose one to schizophrenia, despite the fact that an identical twin of a schizophrenic has about a 50:50 chance of being schizophrenic. (Although the twin has the same DNA, interactions with the environment must play a part.) Because this disease, which affects 0.5%–1% of the world's population, appears to be caused by the interaction of several environmental factors acting upon those with a genetic predisposition, the contribution of each genetic agent has so far been difficult to discern. Consequently, the National Institutes of Health (NIH) has launched a $6 million a year effort to contact additional families to map SNPs that are associated with schizophrenia.[10] One gene that

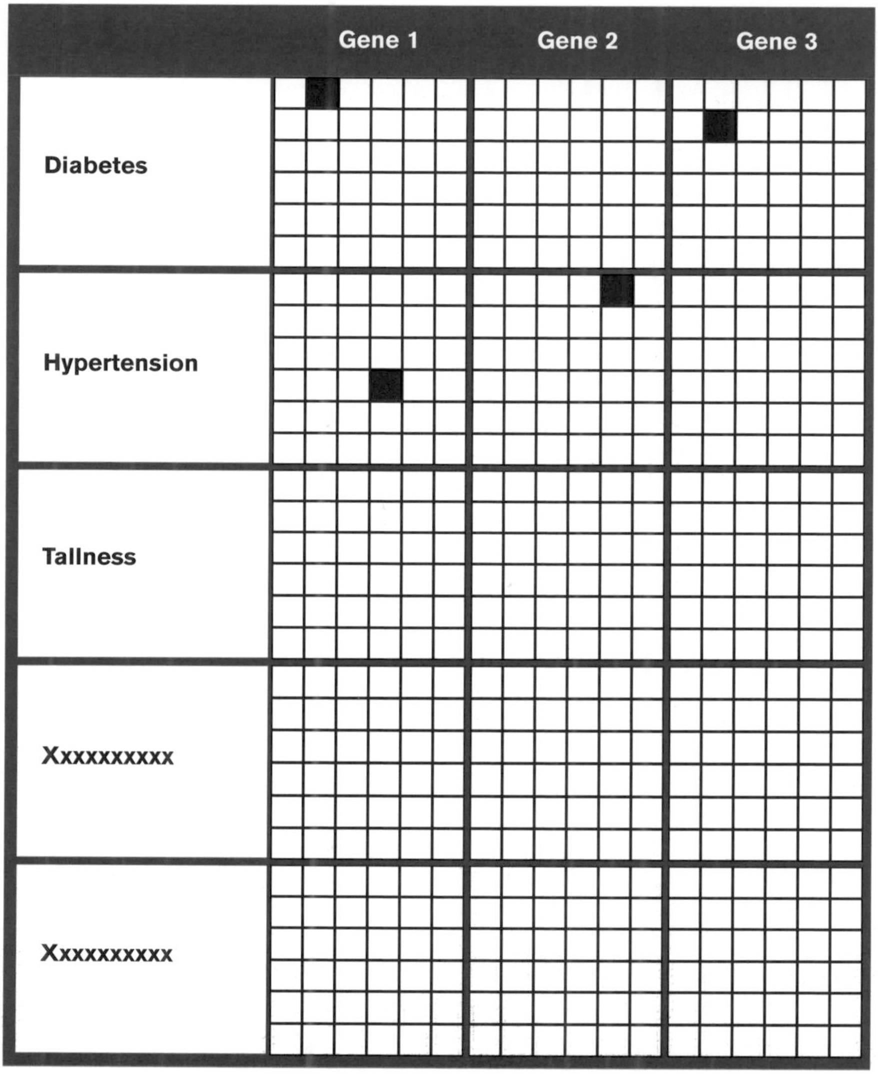

	Gene 1	Gene 2	Gene 3
Diabetes			
Hypertension			
Tallness			
Xxxxxxxxxx			
Xxxxxxxxxx			

Figure 5.1 Shown here is a hypothetical SNP index table that details the genetic variation (for example, an A instead of a T within a gene) that is associated with particular traits. In this way, Lander and his colleagues would like to see what sets of genes are involved in particular traits or genetic disorders.

has caught the attention of the NIH codes for the catechol-O-methyltransferase or COMT gene. The protein coded for by this gene can break down dopamine, a neurotransmitter in the brain and several SNPs within the COMT gene are associated with the onset of schizophrenia. Certain variants created by COMT SNPs can affect the level of this protein in the brain, sometimes causing too much to be made and too much dopamine to be destroyed.[11] Revelations such as these and the use of drugs that target specific neural functions are changing the way that mental illness is viewed. These findings and others that provide biochemical causes for mental illness raise the legitimacy of including mental illnesses among conditions that should be covered by health insurance.

The Human Genome Project not only generated databases, it also spurred the development of several technologies that allow scientists to look at genetic information on a global scale. Companies are beginning to design **microarrays** that can be used as diagnostic tools (Figure 5.2). Microarrays contain thousands

PRO OR CON?

Companies need to have the intellectual property rights to their discoveries in order to make their efforts profitable. But scientists need access to sequence information in order to analyze it, for example, to make comparisons between genes in different living organisms and to plan experiments that make use of those sequences. Do you think that a company should be able to obtain a patent for a novel gene sequence it discovers and then remove that sequence from the database? Why or why not? What would happen if many such human sequences became the private property of a corporation? How can this conflict be resolved so that all can benefit?

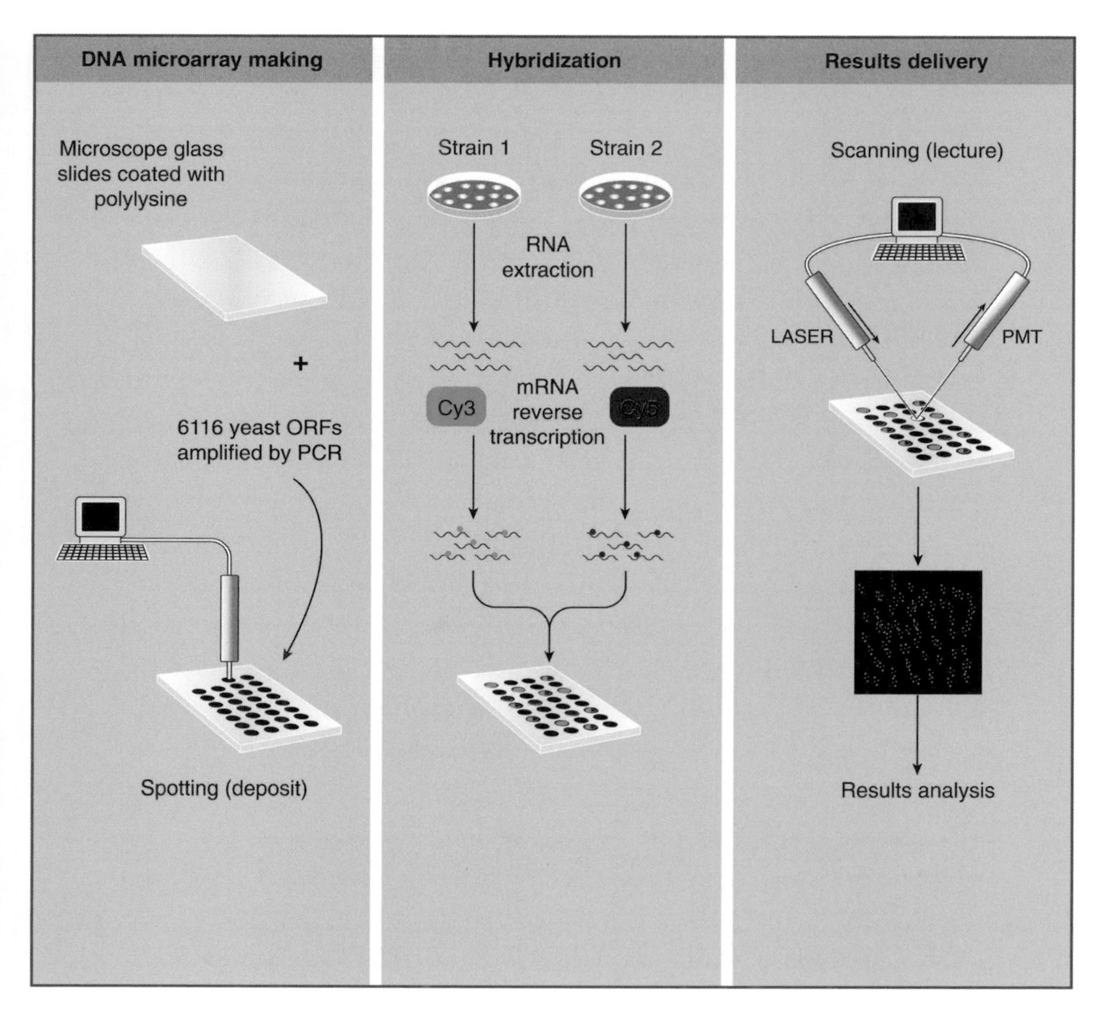

Figure 5.2 DNA microarrays are used to detect differences in mRNA expression. The DNA corresponding to all of the genes in an organism (for example the 6,116 yeast ORFs) are placed, one gene at a time, in spots across a small slide. Then mRNA is isolated from cells whose treatment or condition (such as diseased or normal) is to be compared. The mRNA is converted to labeled DNA, using fluorescent colors to distinguish their origins. The labeled DNA is applied to the DNA on the microarray and attaches to the complementary DNA. The pattern of fluorescent red, green, and yellow (red + green = yellow) spots is read using a photomultiplier tube and computer to indicate which genes' mRNA expression increased, decreased, or remained the same.

of spots of DNA arranged in rows on slides. Sometimes microarrays are arranged so that each spot corresponds to one gene in the genome. For example, the first spot on the slide contains several identical molecules of a piece of DNA that correspond uniquely to gene #1 of an organism. The next spot contain pieces of DNA that correspond to gene #2. The next spot will have a set of DNA molecules that correspond to gene #3. In this way, one slide can contain spots that correspond to each gene in the genome.

To produce the microarrays, you need to have many identical copies of each piece of DNA. The pieces are then loaded onto small slides by robots so that one or a few slides can hold pieces of DNA that represent every gene in the genome. One way to make many copies of a portion of DNA is to use the process called PCR, or the polymerase chain reaction that was described in Chapter 3. For yeast, 6000 PCR reactions, can amplify 6,000 sections of the genome corresponding to 6,000 genes. (The same can be done for the 20,000–25,000 human genes or the genes from any genome using cDNAs that were formed by making copies of isolated mRNAs.) After the PCR reactions have created pieces of DNA using each cDNA as a template, a robot armed with 96 tiny fountain pen-like devices picks up a sample of DNA from each reaction from a 96-well plate and places the samples in rows on a slide. Obviously more than 60 reaction plates are needed to create 6,000 spots of DNA that respond to 6,000 genes on a single slide.

A particular kind of microarray is a **GeneChip®** that chemically builds specific strands of DNA in each spot. If you have ever created a batik in art class, you will recognize the principle behind this masking process. When making a batik, you first place wax over the areas of a piece of cloth to remain uncolored and then dip the entire piece of cloth into a vat of dye. You can then add more wax to cover a portion of the newly colored material and dip the entire piece into a second color. Any unwaxed areas will take on the new color on top of the old. This process can be repeated as often as one wants until most of the batik is covered in wax, except for those areas exposed to all of the colors including the last.

In several ways, creating a GeneChip is similar. Here, instead of dipping a slide into a vat of dye, the slide is covered with a particular base, A, G, C, or T. Each base carries with it another molecule that is a "mask" that will prevent the addition of other bases until that mask is removed. The mask can be removed only by exposure to light. Therefore additional bases will be added to each growing chain only if that chain has "seen the light." By creating grids that shield some bases and not others when the light shines, Dr. Stephen Fodor found a way to simultaneously add bases across the chip. Nonetheless, each position eventually holds pieces of DNA 20 bases long that have a unique sequence. (You can hear Dr. Fodor discuss this process and see animations of GeneChip creation at the DNAi Website at *www.dnai.org.* Click on "Manipulation," then "Techniques," then "Large Scale Analysis" to view the material on GeneChip.)

Microarray and chip design is based on the principle that DNA will bind to its complementary strand. Such arrays can be used in genetic testing to identify particular SNPs within an individual's genome. For example, one can design a chip so that it contains multiple copies of DNA sequence, that correspond to each variation of a particular DNA sequence that is known. In order to test a person's DNA to see which SNP is present, sections of his or her genomic DNA are copied by PCR and labeled with a fluorescent tag. The chip is then flooded with these tagged bits of DNA. The tagged pieces will automatically find their complement and light up that section of the chip, indicating whether or not a particular SNP is present in the DNA. Such chips are being made to diagnose a variety of illnesses. It may not be long until we are routinely screened for a variety of inherited disorders with such chips. This ability to detect a predisposition to illness, before the illness has developed, will raise many ethical, social, and legal issues surrounding diagnostic tools that have the power to create a personal record of disease predisposition. Prominent people such as Dr. Francis Collins, columnist Stanley Crouch, Supreme Court Justice Stephen Breyer, Congresswoman Louis Slaughter, and Nadine Strossen, president of the ACLU, recently discussed many of these issues on the highly entertaining and informative Public Broadcasting System program *Our Genes/Our Choices*, which can be found at www.pbs.org/fredfriendly/ourgenes.

Stop and Consider

How do microarrays help scientists determine the genes that are involved in certain diseases or genetic conditions?

Global Variation in mRNA

In addition to determining the variation present in our DNA and correlating those SNPs and other forms of variation with disease, scientists would also like to monitor the types of mRNA that

Ethics and Genetics

If you have access to an Internet connection, read the essay by Mary Ann Cutter, Ph.D., of the University of Colorado about a way to reach an ethical decision at *http://www.pbs.org/fredfriendly/ourgenes/should_we.html.* Having done that, apply her "Five ways of talking about what's good or right" and "Five steps for coming to an ethical decision" to the question of whether or not one should be tested for genetic predisposition to disease. Listen to the conflicting points of view on this issue from the PBS program *Our Genes/Our Choices* at *www.pbs.org/fredfriendly/ourgenes* under "Who Gets to Know? Genetics and Privacy." Listen as Congresswoman Louise Slaughter and Supreme Court Justice Stephen Breyer wrestle with this issue. Discuss the ethics of releasing such diagnostic genetic information to health insurance providers. Why might the perspectives of the Congresswoman and the Justice be different? What role should the government play to ensure justice in issues of privacy and healthcare?

we make as a way to diagnose disease and monitor treatment. Microarrays and chips provide a great way to do this. The yeast community (scientists who have devoted their lives to the study of cellular processes in yeast) was one of the first to organize itself to determine globally what was going on inside of yeast. Since yeast cells have only about 6,000 genes, it was relatively easy to make microarrays with all 6,000 yeast genes. Not only could these slides identify which genes are being read to make mRNA, they could also allow researchers to test which mRNAs were made by yeast under a particular condition. For example, yeast scientists at Stanford and at the University of California at San Francisco were interested in which genes were turned on (made mRNA) in response to a reproductive process in yeast called **sporulation**[12] when a yeast cell morphs into four spores. During this process, the cell distributes its chromosomes so that each spore gets only half the original number. But each spore must get an intact set, not a random mix. (Ultimately, cells derived from spores may mate just as eggs and sperm combine in other organisms like us.) In the usual type of cell division, called **mitosis**, cells duplicate their DNA and then divide so that each new cell has the same DNA as the cell that divided. But during **meiosis**, a different pattern of division takes place so that the spores have only half the information and number of chromosomes.

The yeast scientists knew that in order to have yeast chromosomes line up the way ours do during meiosis, just before we make eggs or sperm, and in order to change their morphology from a single cell into four little spores inside of a sac, the yeast cell needs to change the products made by its genes. It's as if the yeast were a factory gearing up for their Christmas season when new toys are made. Scientists want to know which genes are being read, which are not, and which have changed as the yeast tooled up for sporulation. In order to detect not only which genes are turned *on* but which genes are turned *off*, the scientists used

PCR to make large amounts of DNA from mRNAs produced by the sporulating yeast and in the process attached special tags that appear green under fluorescent light. Similarly, the scientists made fluorescent red-tagged DNA from mRNA produced by yeast that were not sporulating. Then the scientists poured the tagged DNAs over yeast microarrays that already contained DNAs corresponding to all 6,000 yeast genes. DNA generated from the mRNAs of both sporulating and non-sporulating cells can bind in the same spot if both cell types are making the same mRNA. Where the two types of mRNA bind in one section, that spot appears yellow (red + green). However, if more mRNA is made by the sporulating cell, the corresponding spot on the array is green. If more mRNA is made by the non-sporulating cell, the corresponding spot is red. Therefore, depending upon whether the square is red, yellow, or green or shades in between, scientists can tell whether sporulation turns mRNA production from a particular gene up, down, leaves it at its ordinary level, or never turns it on at all.

The scientists do not make these determinations only once during the time the yeast cell is going through sporulation. They make the same kinds of measurements at several different intervals after cells are triggered to sporulate by being placed in nitrogen-deficient medium. They determine what each particular gene is doing over time, whether it is being turned up, turned down, or never turned on at all. Because the responses of 6,000 genes creates an enormous amount of data, the scientists present that data in grids containing colorful rows of changing colors for each gene. This is a great improvement over tracking the activity of only one gene at a time. The convention now is to report increased mRNA production for the experimental condition as a red spot, a decrease in mRNA production as a green spot, and no change as a black spot (instead of yellow). Although the data for the yeast sporulation study was collected as described above, it was reported in *Science* according to this new color protocol (Figure 5.3).

Once scientists learned how to do this with yeast, in no time at all the variety of microarrays and chips expanded. After other genomes were sequenced or after scientists had accumulated enough cDNAs to create the microarrays, they began to test the global production of mRNA in other organisms. At first they were interested in the mRNA output of specific tissues within organisms, such as brain cells or skin cells, and eventually they realized that they could measure the global production of mRNA in cancer cells. Not only could they measure the global production of mRNA, but they could see if there were differences that they had missed before in cancer cells that otherwise looked the same as their normal counterparts.

For example, a study attempted to sort breast cancers based upon their individual global mRNA expression patterns.[13] At first glance, the patterns obtained from 42 cancer patients seemed incredibly varied. However data from a second analysis of some of those tumors allowed the researchers to pick up the expression of particular genes that did not change when the tumors were retested. (Unfortunately, this data also showed that the chemotherapy was largely not effective in changing the global mRNA expression patterns of these cells.) Because of the retesting data, they could select a subset of about 500 genes that showed greater variation among tumors from different patients than between the resampled tumors from the same patient. When the computer sorted the tumors according to this "intrinsic gene subset," the tumors clearly fell into two subtypes that largely correlated with their expression of the estrogen receptor. The presence or absence of the estrogen receptor protein on breast cancer cells has been a useful way to determine the choice of treatment for a number of years. The challenge now is how to turn this tabulation of gene expression into targeted therapies that return the gene expression pattern to a normal one.

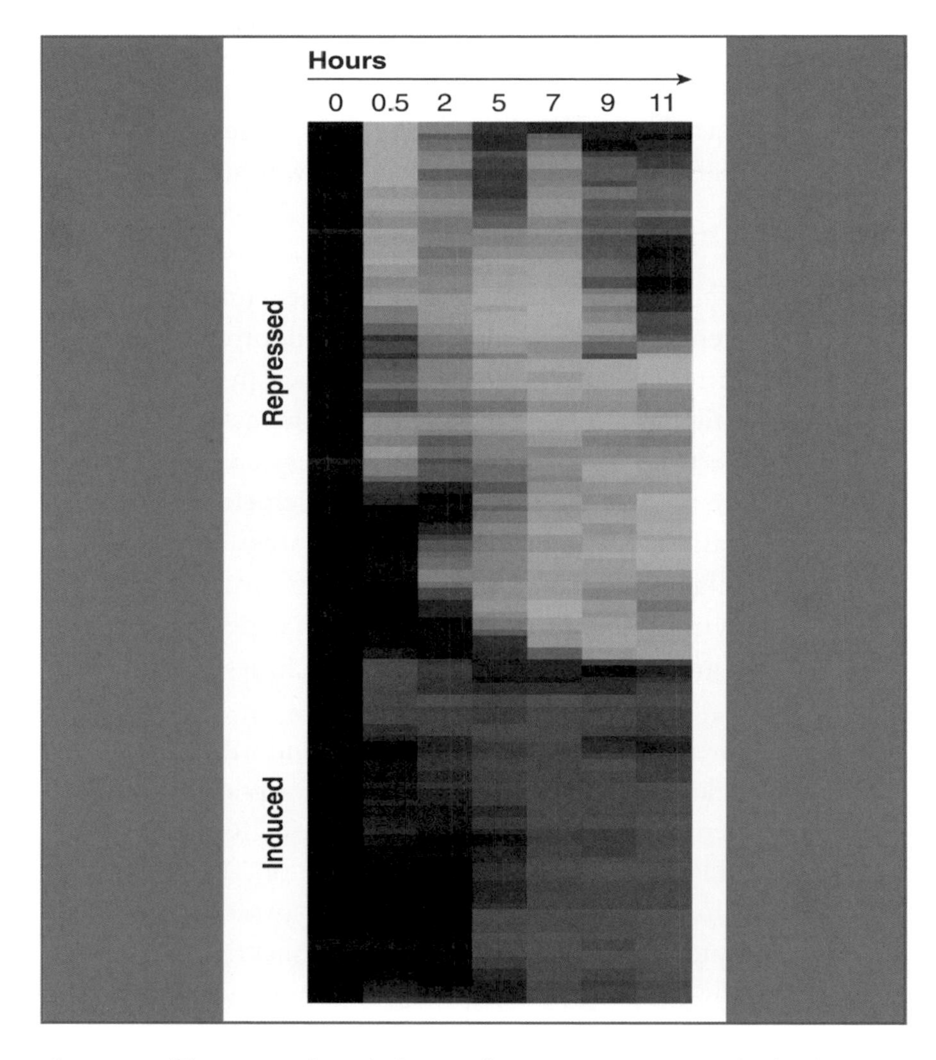

Figure 5.3 The expression of almost all yeast genes was tracked using microarrays at various times after yeast were starved for nitrogen. This treatment induced sporulation, the reproduction of yeast cells. About 1,000 genes showed significant changes in expression. Here, each row represents the expression pattern of a different gene and each column is a successively later time point. Red indicates an increase in regulation (up-regulation) and green indicates a decrease in regulation (down-regulation). *Source: S. Chu, et al., "The Transcription Program of Sporulation in Budding Yeast,"* Science *282 (1998): 699–705.*

This points out one of the challenges about the development of global analysis methods in biology and medicine. So much information is collected, often in the absence of a precisely formulated hypothesis, that it can be difficult to know how to use the data.

Proteomics

Scientists optimistically think that when they understand not only which genes, mRNAs, and protein structures distinguish cells under different conditions, but also have insight into the ways that particular proteins interact with one another, we will begin to understand the fundamental processes of life so well that we will be able to precisely diagnose which elements of this system are malfunctioning in disease. Ultimately, we will know which particular proteins within networks to target with drugs and which environmental factors to change in order to inactivate proteins run amok or enhance protein machines operating at half-speed.

The next step is a big one. Because DNA and mRNA are fairly simple molecules, we know a lot about DNA variation and mRNA expression, even on a global scale. We now have PCR machines that can deliver millions of molecules of any 1,000 or 2,000 base pairs of DNA we choose in just a few hours. But proteins are much more complicated. We are beginning to learn how to take proteins apart efficiently as a way of identifying their primary amino acid sequences, but chemically putting them together or determining the location of each atom within a particular protein machine is still an arduous process. In the past, it was necessary for a laboratory to clone a gene for a protein, express great quantities of that protein usually in bacteria, isolate and purify the protein, make regular crystals of the protein (the hardest part), and then use x-rays to map the intricate location of each and every atom within the protein. A newer technique called nuclear magnetic resonance or NMR yields data more easily, but not with the resolution of

X-ray crystallography. All of the protein data from labs around the world is deposited in the Protein Data Bank, *http://pdbbeta.rcsb.org*. Other groups have taken upon themselves the task of annotating each protein, scanning its predicted sequence for clues about 3-D structure and recording its structural modules, its functional modules, and its expected or known location within the cell. Still others are using a technique based upon **mass spectrometry** and the charge and size of proteins to determine the identity of *all* of the proteins in a sample. We are just beginning to analyze ovarian cancers in this way in an effort to find protein differences in the blood of people who either do not have ovarian cancer or have progressed to various stages of the disease. The results have been quite promising and are posted at the National Cancer Institute's proteomics Website at *http://home.ccr.cancer.gov/ncifdaproteomics/*. The real test of proteomics, this focus on proteins, is how these proteins interact with other proteins, nucleic acids, and other networks of molecules within cells.

The Website *http://www.seoulin.co.kr/Up/* provided by the Korean biotechnology company Seoulin illustrates some of these complex interactions with tremendous visual impact. Its "Virtual Biology" page animates the intricate dance of the *period*, *timeless*, *cycle*, and *clock* proteins made by the fruit fly *Drosophila*. These proteins regulate the fly's daily activity according to a molecular clock called a **circadian clock**. These four proteins must be made at the right time (and some regularly degraded) and they must be regularly escorted back into the nucleus to work their magic on DNA transcription. This Website also animates the intricate series of molecular interactions that mediate apoptosis, the program by which a cell commits suicide. **Apoptosis**, or programmed cell death, is responsible for many steps in the development of our bodies including the disappearance of the webbing that we initially have between our fingers as a fetus. Although the shear number of interactions shown in these animations is staggering, the many more interacting players and transient binding events there must be in the interlocking networks are not shown. Many Websites have put together information gleaned from a variety of sources to illustrate the myriad ways in which proteins interact to affect the processes of life. For example, one can search the Biomolecular Interaction Network Database or BIND Website at *http://bind.ca/index.jsp?pg=0* for pathways, molecular complexes, and individual interactions associated with a particular gene. A powerful example of network interactions is seen in cancer cells that have stripped the brakes on cell division and souped up the activators of cell division. The drug called Gleevec distributed by Novartis stops such an activator dead in its tracks.[14] Gleevec works by stopping an overactive enzyme protein that has been wreaking havoc by stimulating the production of too many white blood cells in a person with chronic myeloid leukemia (CML). The expression of this protein is turned on by the fusion of pieces of two chromosomes. The specificity of Gleevec is elegant and is the ideal of personal pharmacology of the future.

Stuart Schreiber, who gained tremendous insights from watching the male cockroaches respond to his love potion, would like to make small molecules that effectively throw a monkey wrench into defective proteins and stop them cold. He has generated tens of thousands of such randomly shaped chemicals and is testing what works in a kind of natural selection, by pouring the molecules over their target, the defective protein (again affixed to slides), and seeing what sticks. Then he tests their show-stopping effect in live

Gleevec

You can hear from the first Gleevec recipient, "Bud," who talks about his chronic myeloid leukemia treatment at the DNAi Website at *www.dnai.org*. Select, in turn, "Applications," "Genes and Medicine," and then "Drug Design." Bud considers himself lucky to be alive.

cells. In this way, he believes that he will identify chemicals that will neutralize defective proteins that are gumming up the works, chemicals that can be turned into useful drugs.

Although we only recently entered the genomic age, in many ways it is already behind us. The challenge of the future is in proteomics, the study of the structure, function, location, and interacting networks of these cellular protein machines. This challenge suggests why an interdisciplinary approach to biology will also be the wave of the future.

CONNECTIONS

We began this chapter by looking at the structure of the BRCT domains of BRCA1 and their binding site for the DNA helicase required to repair DNA. Some of the nucleotide changes in the DNA sequence that encodes that binding area lead to heritable early-onset breast cancer. As an extension of the Human Genome Project, scientists would like to correlate the location of all SNPs with disease, especially in an effort to learn about diseases for which there are many contributing genetic mutations.

Also as a result of the Human Genome Project, we have developed several high-throughput technologies such as microarrays and chips that allow us to scan a genome for SNPs. We can also use these arrays to determine the entire mRNA output of a cell and attempt to use such information to enhance diagnoses and follow the efficacy of treatments. Finally, we are beginning the age of proteomics in which we hope to personalize medicine and tailor treatment, especially for cancer, so as to provide a specific and hopefully globally non-toxic intervention.

Throughout all of this exploration, we must be humbled by nature's laboratory notebooks that, as Dr. Eric Lander noted, have been laid open before us.[15] Almost 3 billion years of creatures living out their lives and determining those of future generations as they blindly pass on their DNA have brought us to today's array of organisms. The successes and failures of each are recorded in their DNA, records which we are just beginning to read. As with all such powerful tools, it is up to us to use them wisely.

FOR MORE INFORMATION

For more information about the concepts discussed in this chapter, explore the following Websites:

To explore genetic tests in general and the BRCA1 mutation test, go to *http://www.dnai.org/index.htm* and click on the tab "Applications." From the page that comes up, select the tab "Genes and Medicine" to learn about the breast cancer gene BRCA1. Under the section "gene hunting," listen to Dr. Mary Claire King who discusses her 20 years of looking for the BRCA1 gene. Watch the video as she explains how she developed pedigrees of the affected families and looked for genetic markers that were linked to the BRCA1 gene. There is also an animation that explains the concept of genetic markers. Under the section "gene testing," listen to Dr. Barbara Weber, the director of the Breast Cancer Center Program at the University of Pennsylvania, explain her work to isolate a breast cancer gene. Her patient was a woman named Vicky who had many relatives with breast cancer and who herself tested positive. Be sure to listen as Denise, Vicky's sister, talks about how she felt after finding out the results of her test for early-onset breast cancer.

You can learn more about Hutchison-Gilford Progeria research in an NIH News Release at *http://www.genome.gov/11006962* and in the Online Mendelian Inheritance in Man file #176670 at *http://www.ncbi.nlm.nih.gov/entrez/dispomim.cgi?id=176670.* Notice how many other diseases are associated with defects in the Lamin A/C gene.

Go to the *www.dnai.org* Website to see how microarrays are put together. Hear from Dr. Patrick Brown of Stanford University, who developed the microarray, explain his version of this process.

Using the NCBI Server to Learn About a Protein

This chapter and the next will take you into the databases of genomes and proteins that researchers have assembled, so that you can do what scientists do each day in the lab: uncover a particular protein in all those millions of records, see how the protein is shared among many different creatures, and discover what changes from organism to organism. You can investigate how a little mistake here and there in the protein can have devastating effects on people's lives. In this chapter, we will use the databases to learn more about the bacterial potassium channel KcsA and its relevance to children with epilepsy. For some of this chapter you will need a computer with access to the Internet.

Dr. Roderick MacKinnon of Rockefeller University shared the Nobel Prize in chemistry in 2003 for determining the exact position of the atoms that fashion the bacterial potassium channel KcsA.[1] He did this using a well-established technique called x-ray crystallography. The hardest part of this technique is making crystals of protein. Dr. MacKinnon put the gene for the KcsA channel from a bacterium called *Streptomyces lividans* into another bacterium called *Escherichia coli.* As the *E. coli* multiplied, they also produced the KcsA channel. Dr. MacKinnon and the members of his laboratory isolated that protein by breaking open the bacteria and dissolving much of the fatty membrane where the channel was located. After doing this to several large cultures of the bacteria and isolating and concentrating the channel protein, he tried many different conditions to coax the pure protein to form crystals, just like crystals of salt that you shake on your scrambled eggs. Once he had a crystal that was large enough, he shined an x-ray beam through the crystal. Because the crystal was a regular and ordered arrangement of the protein molecules, the patterns made by the x-rays as they were blocked or deflected by atoms within the crystal could identify the location of each atom within the protein.

Experiments performed much earlier had helped scientists form a mental picture about the structure of such a channel—before they ever saw one. Because of the speed at which potassium ions could pass through the channel, scientists assumed that the channel's narrowest part, the filter, must be very short compared to the rest of it. (A long filter would significantly slow the passage of ions.) After determining the sequence for potassium channels from many different organisms, scientists noticed that all of the sequences coded for an almost invariant string of four amino acids within the channel protein: glycine, **tyrosine**, glycine, and **aspartate** or **GYGD** for short (Figure 6.1). They already knew that tinkering with this sequence would slow potassium ions as they moved through the channel and reduce the channel's preference for potassium over sodium. Therefore, they knew that this short string of amino acids had some important role in the selectivity filter, the filter that allows only one sodium ion to pass through for every 10,000 potassium ions that streak through.

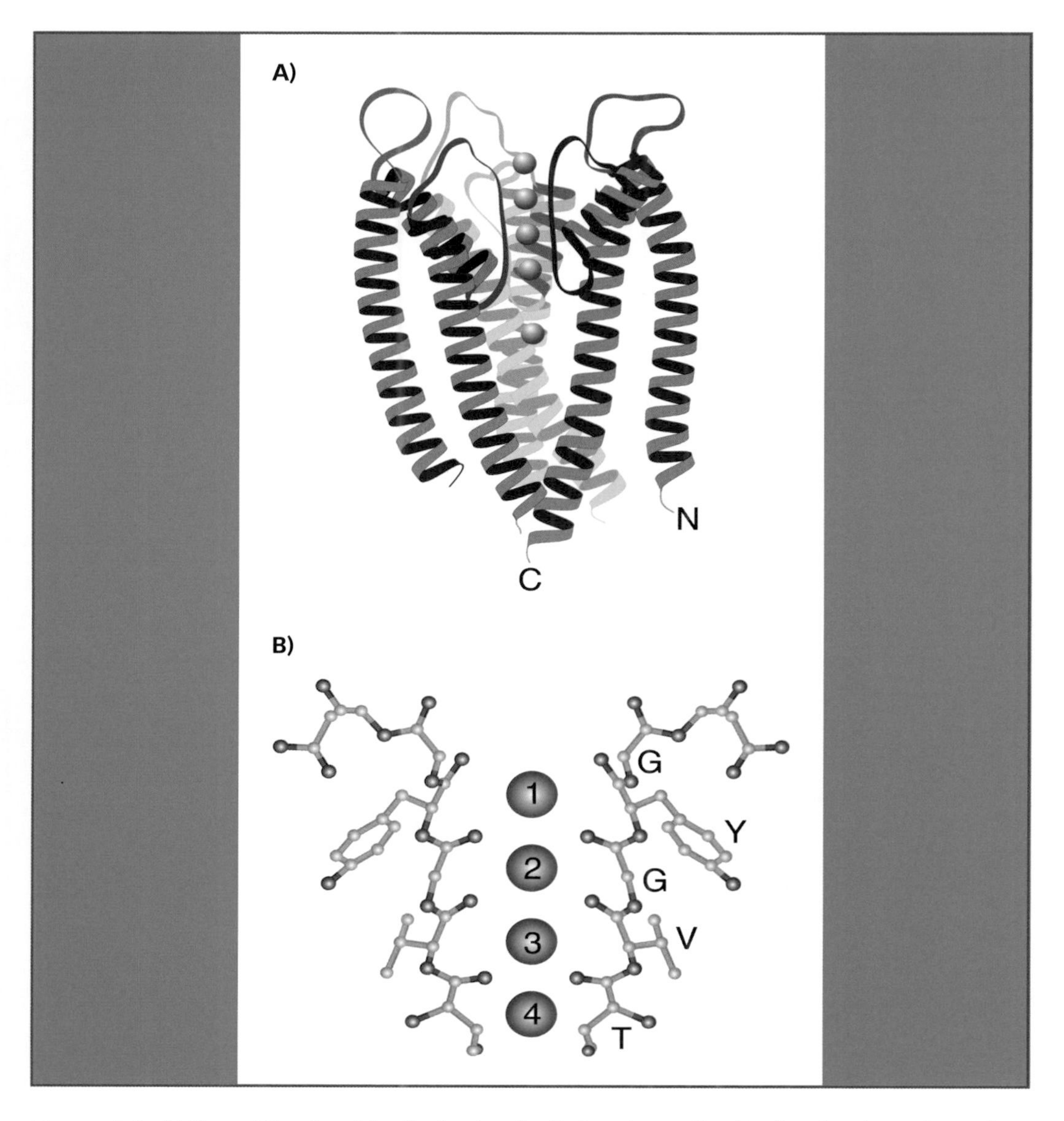

Figure 6.1 A) Two of the four identical subunits that make up the KcsA potassium channel are shown here. Each subunit contains three alpha helical sections, indicated by the coiled ribbons. The position of five green potassium ions are shown as they move through the channel and its narrow selectivity filter. (B) The invariant GYG portion is shown in this close-up of a potassium channel's selectivity filter. (Only two of the four subunits are shown). *Source: R. MacKinnon, "Nobel Lecture. Potassium Channels and the Atomic Basis of Selective Ion Conduction,"* Bioscience Reports *24 (2004): 75–100.*

The crystal structure showed why these particular amino acids were important. These amino acids include key **residues** that line the narrow passageway and reach out with their carbonyl oxygens to create the selectivity filter. Potassium channels from many species, from bacteria to humans, have essentially the same inverted tepee that contains the narrow selectivity filter. However, many of these channels are built from significantly longer proteins. In this chapter, you will learn how to use the National Center for Biotechnology Information (NCBI) server to access information about a protein and the gene that codes for that protein. You will learn how to fish the NCBI databases for DNA sequences that encode similar proteins or a protein that has a part similar to your protein of interest. As an example, we will discuss how to look for channels or parts of channels in other organisms using the protein sequence for the bacterial KcsA potassium channel as bait.

Much of the base sequence and protein data are available through the NCBI server. The information available there or through other servers around the world about a particular gene or protein are almost endless. Here's a partial list of what can be found in the databases: (1) the mRNAs and proteins known to be made or predicted to be made from that stretch of nucleotides; (2) references to experimental evidence that these mRNAs and proteins exist; (3) regulatory sequences at the beginning and end of genes that control when and where the gene is transcribed; (4) common mutations and their consequences in **model organisms** or in human disease; (5) protein domains that are present; (6) similarities to proteins or protein domains in other organisms and their **evolutionary trees**; (7) the interactions of those proteins with other proteins or with DNA or RNA; (8) a variety of ways in which individual proteins can be modified after they are made; and (9) any other distinguishing features such as the signature sequence for the selectivity filter in a potassium channel. Whew! And that is just a partial list.

As a way of illustrating the power of the databases available through the NCBI, in this chapter we will show you how to answer the kinds of questions posed in Figures 6.2 and 6.3. These questions are all included within two broader questions:

The NCBI Server and UnitProt/Swiss-Prot

Individual scientists and groups of scientists have determined the nucleotide sequences of both individual genes and whole genomes. These sequences belonging to organisms from bacteria to humans have been deposited in databases throughout the world. These include the database Genbank in the United States, the DNA Data Bank of Japan, and the European Molecular Biology Database. Scientists throughout the world can contribute sequence information to any one of these databases, but the information is shared among all three databases on a daily basis. In the United States, these data are available through a central processing center, the National Center for Biotechnology Information (NCBI) at *http://www.ncbi.nlm.nih.gov/*. This center was begun in 1988 to serve as a central hub for a wealth of emerging molecular biology information and has since expanded to offer a variety of tools for organizing and searching for information related to genes and proteins. Since the NCBI was initiated under the auspices of the U.S. National Institutes of Health and the National Library of Medicine, there is a focus on organizing information so that researchers can easily access genetic and proteomic information related to particular diseases.

Some Websites have tried even harder to make information easy to use. The Website UnitProt/Swiss-Prot focuses on proteins. The **curators** at **UnitProt/Swiss-Prot** predict protein sequences based upon the nucleotide sequence that has been deposited and combine this information with data regarding in which cells the gene's DNA has been transcribed into RNA and the proteins made. This Website also **annotates**, or describes, special features of proteins in addition to just giving the nucleotide sequence for a gene, the tissue, or organism the DNA sequence came from and the bibliographical references to its appearance in the database. UnitProt/Swiss-Prot is a useful protein annotation Website that is maintained by the Swiss Institute of Bioinformatics and the European Bioinformatics Institute. Much of their analysis is available through the NCBI server.

1. How can we use the NCBI to find out information about the potassium channel described by MacKinnon's work?

2. How can we use the NCBI to find out about sequences that might code for entire potassium channels or parts of potassium channels in all other organisms?

Figures 6.2 and 6.3 outline the particular questions about the KcsA potassium channel that can be answered at each stage of our computer exploration. Make a mental note that questions of a similar kind can be answered about any protein by following the same set of computer steps.

LEARNING ABOUT KcsA FROM A PROTEIN FILE

One may enter the NCBI Website through *http://www.ncbi.nlm.nih.gov/* which is the home for the National Center for Biotechnology Information (Figure 6.4). In order to learn about any protein, one would adjust the pull-down window to read "Protein" and type in the name of the protein one wishes to study. For example, one could type in the words "KcsA K channel" in the open window to indicate the potassium channel studied by Dr. MacKinnon's group. Selecting "Go" will bring up a series of entries related to our quest. To find the protein purified from the bacterium *Streptomyces lividans* whose structure was determined by Dr. MacKinnon, pick the choice labeled P0A334. This choice will pull up a window with a concise but comprehensive description of the bacterial KcsA potassium channel annotated by the curators at the Swiss-Prot database. Portions of this window are shown in Figure 6.4.

The top of any protein file is labeled with a particular number. For example, the number P0A334 identifies the KcsA channel protein file. The top line of such a file will indicate the number of amino acids in this particular protein; for KcsA, the protein

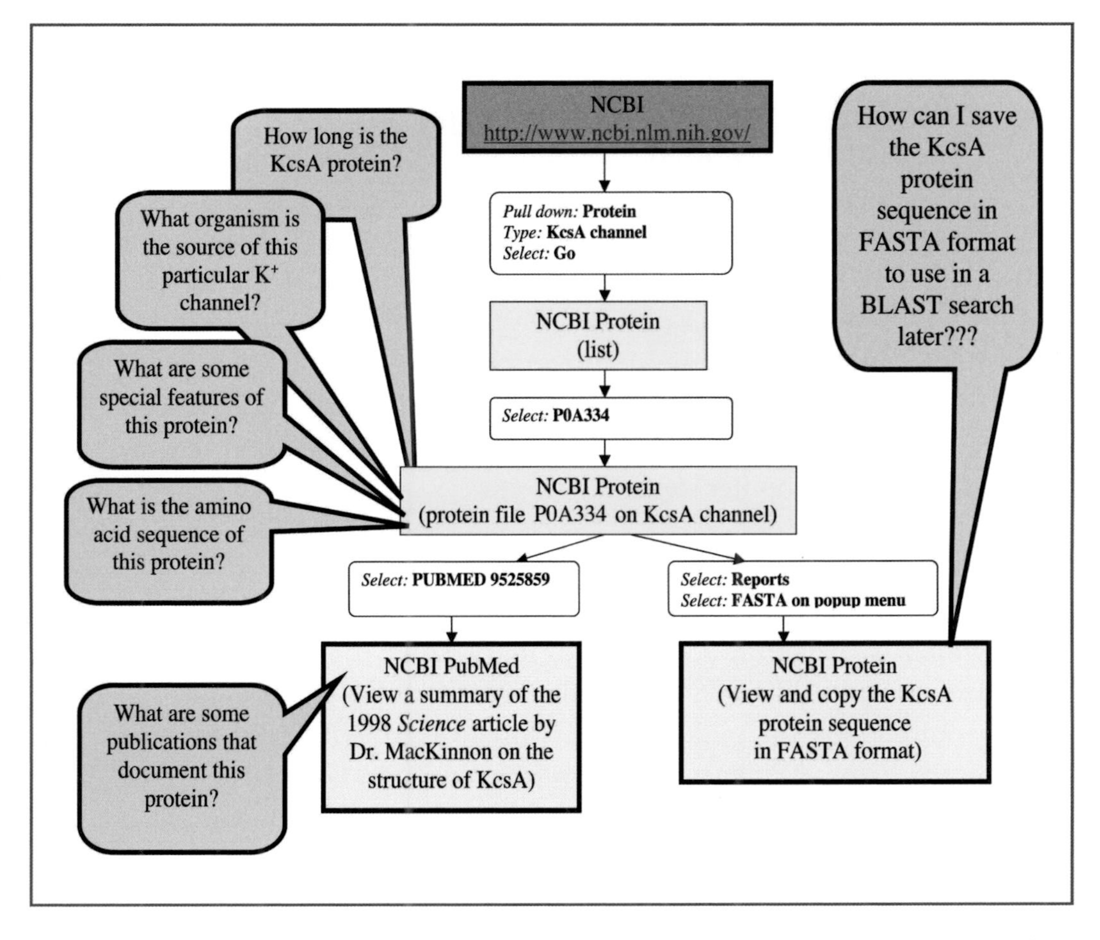

Figure 6.2 You can learn more about the KcsA potassium channel by using the NCBI Website. Follow the flow chart and description in the text to learn about the KcsA potassium channel from Protein file P0A334. (Note that "0" is a zero.) The call-out bubbles indicate the kinds of questions that can be answered by following this computer path.

is 160 amino acids long. The number GI:61226909 was assigned to a portion of DNA sequence that contains the code for this channel, based upon the order in which it was deposited. Other numbers listed under cross-references or "xrefs" refer to additional

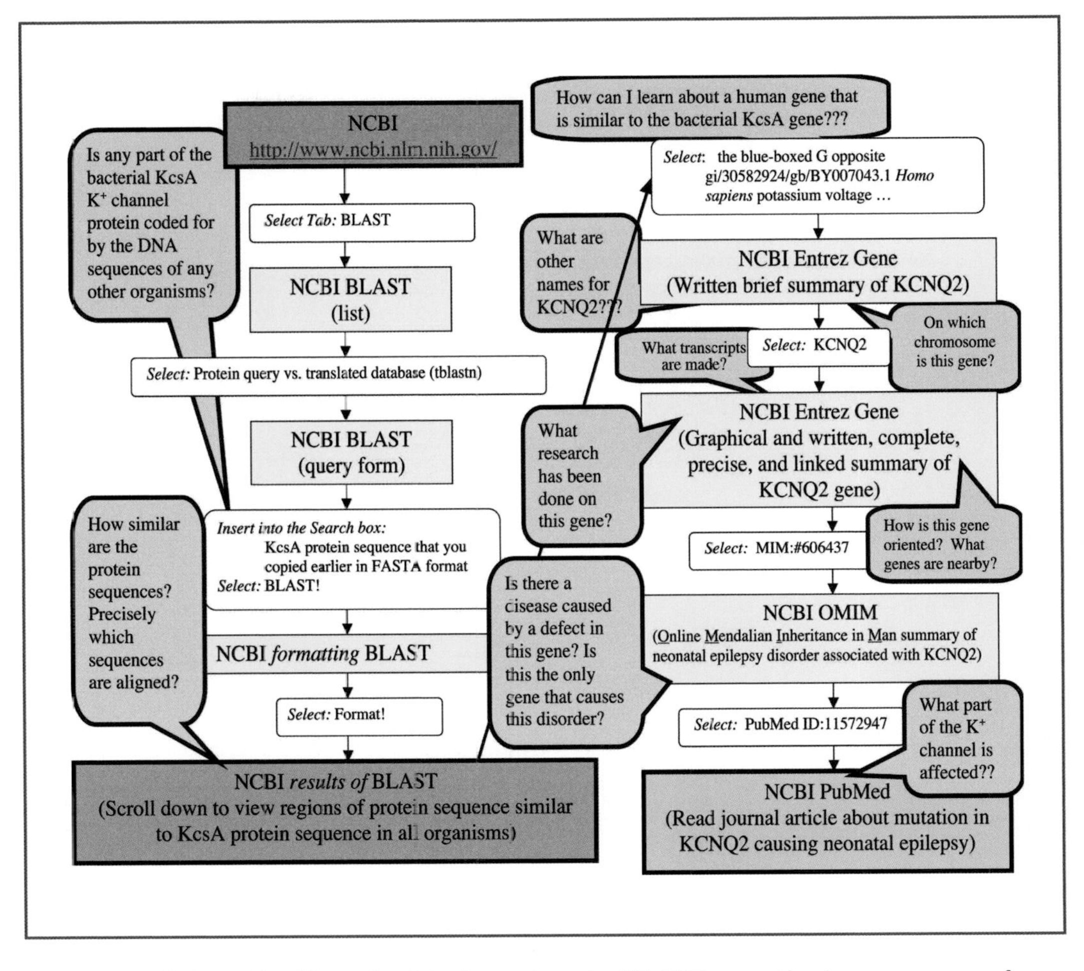

Figure 6.3 Follow the flow chart to learn how to BLAST a particular sequence of DNA (look for sequences like it) and how to access the Entrez Gene page to learn more about a gene. The call-out bubbles indicate the kinds of questions that can be answered by following this computer path.

nucleotide or protein sequence entries or other information that pertains to this protein.

The "Source Organism" label on this page will be followed by the organism from which this channel was identified. For KcsA,

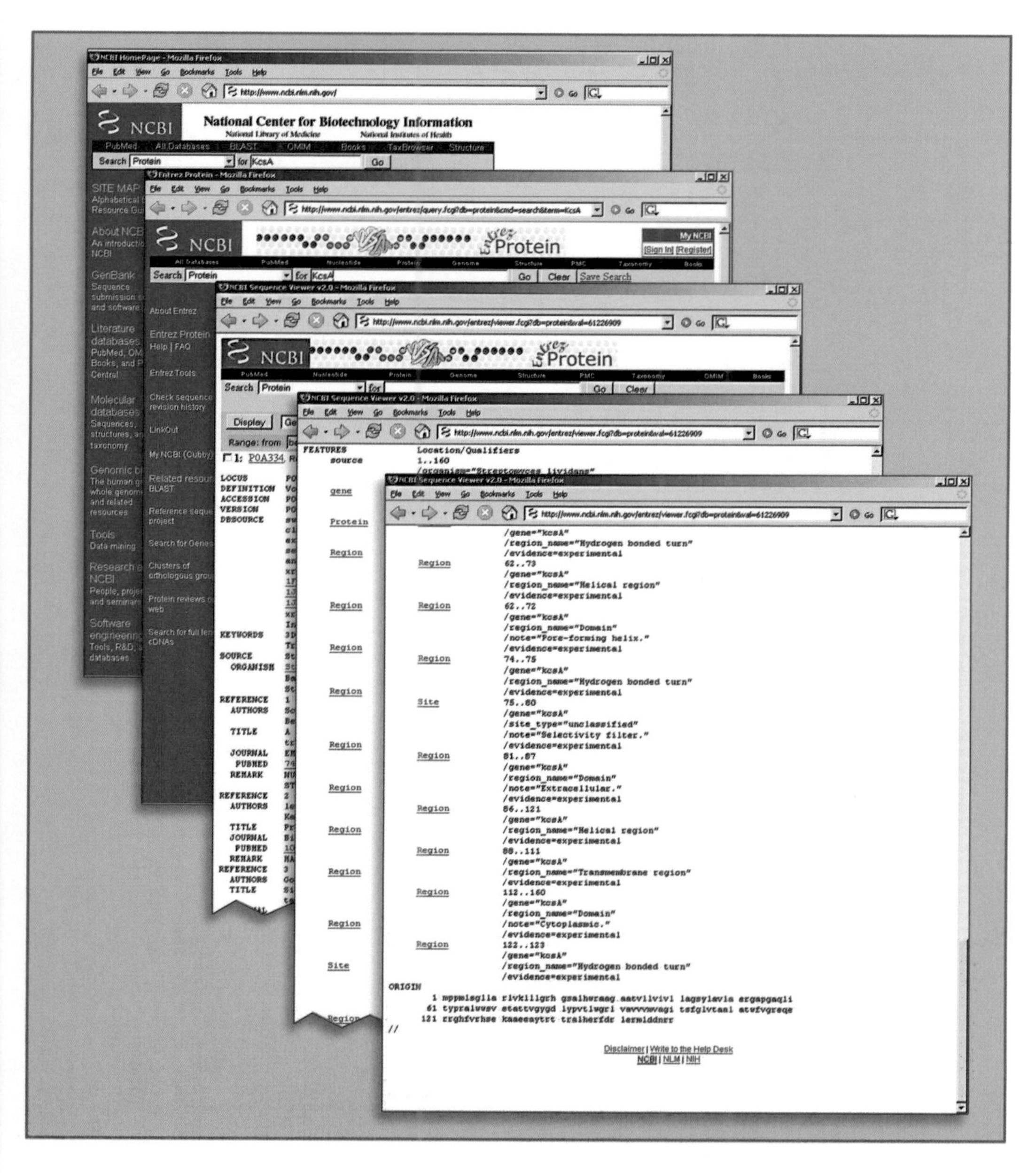

Figure 6.4 The Webpage path from the NCBI Homepage to the bottom of the protein file P0A334 is shown here and described in the text. The amino acid sequence of the KcsA protein is given in lower case letters at the bottom of the file.

the source organism is the bacterium *Streptomyces lividans* that Dr. MacKinnon's laboratory used to make the crystal structure of this KcsA channel. As one scrolls down any such protein page, one will notice that bibliographical references document both the nucleotide sequences and the protein crystal structures. These references are given as **PUBMED** references, a system used by the National Library of Medicine. For example PUBMED reference 9525859 would link to a summary or abstract from the 1998 *Science* article in which Dr. Roderick MacKinnon and his coworkers described the structure of a potassium channel.[2]

Further down the protein file one will find a section labeled "COMMENT" that contains several subheadings. **[FUNCTION]** indicates in what way this protein is used within the cell. **[SUBUNIT]** tells whether or not this protein is made up of subunits and may use the term "homo-" to indicate all the subunits are the same. (If the protein units were different from one another, the term "hetero-" is used.) For example the term "**homotetramer**" would indicate that the protein is made up of four identical subunits, as is true for the potassium channel KcsA. The heading **[SUBCELLULAR LOCATION]** will indicate where within the cell this particular protein is found. For example, for KcsA the subcellular location would be "**integral membrane protein**." We know that the KcsA protein is found within the plasma membrane of this bacterium. But such a designation could indicate any one of a number of internal bags for cells such as the mitochondrial membrane. Another heading is **[MISCELLANEOUS]**. Here we will find out additional information about our favorite protein that has not already been covered. For example, this is where special information about particular regions within a protein might be listed. For the KcsA protein, the comment says, "amino acids 62–79 figure prominently in determining the channel's structure and properties." The purpose of all of these headings is to highlight some of the most informative information about a protein, so that you can learn as much as possible with only a small effort. Having a standard format for the information makes it easier for users to find what they want to know.

The heading **[SIMILARITY]** will let us know whether or not our protein belongs to an identified family of proteins that have a similar structure, which often indicates a similar function. For the KcsA channel, this section tells us that KcsA is part of a family of potassium channels. This protein page also highlights information about particular sections of a protein under the heading **[FEATURES]**. For example, under this section the amino acids 75–80 within the KcsA channel are labeled "selectivity filter" and the heading "experimental evidence" indicates that scientists have conducted experiments to determine that in fact these are the critical amino acids.

At the bottom of each protein file, the actual amino acid sequence for that particular protein is listed. The sequence is given using the one-letter code that corresponds to each amino acid. For example, the code for glycine is G, leucine is L, and alanine is A. For many amino acids, the corresponding amino acid is clear, but for some such as lysine, which is K, one must just memorize this code or have it handy for reference. The code was first discussed in Chapter 2 and is shown in Figure 2.2.

USING BLAST TO SEARCH FOR SIMILAR PROTEINS

In order to search for proteins that are similar, at least in part, to another protein, one may use another option available through the NCBI server called BLAST.[3] Here one may compare any protein or nucleotide sequence with all of the other protein or nucleotide sequences that have been deposited by scientists around the world in GenBank in the United States, the DNA Data Bank of Japan, the European Molecular Biology Database, and the Protein Data Bank. These data are regularly shared so that GenBank has all of the data collected from around the world. However, it is good to remember that there can be delays in processing contributed information and some preliminary data is excluded from the searchable databases. To search the **nr**, or **nonredundant**, database at GenBank, one must first copy the protein sequence in a special format called **FASTA format**. A FASTA formatted sequence has some initial identifiers and then the entire protein sequence represented as rows of single letters that stand for amino acids. The protein page that describes a particular protein, just like protein file P0A334 described above, has a pull-down menu labeled "Reports" that brings up a FASTA formatted sequence of the protein that one may copy and paste into a search.

For example, one may use BLAST to ask the question, "Is any part of the bacterial KcsA K^+ channel protein coded for by the DNA sequences of any other organism?" BLAST may be accessed via a tab at the top of the NCBI page. Although there are several ways to use BLAST to ask this question, one of the best ways is to use an option called **tblastn** which is also labeled **Protein query vs. translated database**. This asks the computer to look for base sequences that can be translated into amino acid sequences that are similar to at least a portion of your protein.

The matches to your protein sequence appear in three forms. The first is a series of lines. The topmost line represents your sequence. All of the lines underneath represent individual sequences that are similar to your sequence. Lines that are as long as the top

The Potassium Channel Revisited

Take another look at the potassium channel illustration on page 16. The middle image shows a cutaway view of the bacterial potassium channel known as KcsA and the probable location of potassium ions as they move through. One can see from the diagram at the bottom of the page that this channel is made up of four identical proteins that are represented by four colored ribbons.

line represent protein sequences that have some similarity across the entire protein. Other lines may be shorter. The proteins with shorter lines match or are similar to only a portion of your sequence. The color of the line gives you an indication of how close the match is. It is easy to see that a change from alanine to valine when comparing one protein to another is not a big change. Both of these amino acids are hydrophobic and relatively small. In contrast, a change from alanine, which has no charge on its side-chain, to lysine, which has a positive charge on its side-chain, would indicate a much bigger difference.

BLAST does not require every amino acid to match or even be similar for a region to be considered to be similar. The rules for how these matches are decided are beyond the scope of this book, but a number is assigned to each amino acid to indicate how similar that one is to the corresponding amino acid in the sequence you put in. Points are assigned so that BLAST determines whether or not a region of a protein could be as similar as it is by chance alone. In addition, there are penalties for gaps where the computer must jump a few amino acids in order to find the next region that is similar.

The second way that the matches to your protein sequence appear is as a list of the sequences that were indicated above by lines. Each line gives an **accession number** and some information about the sequence. There are several live links in these sequences to individual files about these sequences and to the alignment with your protein that is shown in the third section of this report. There are also two scores associated with each line. One is an **E value** or "**Expect value**," a measure that indicates the likelihood of finding such a similar sequence by chance. The significance of each match is roughly indicated by the "E value." The smaller the value is, the better the match. For some scientists, an E value for the result of a BLAST search of 10^{-3} or smaller suggests that these sequences might have had a common ancestor in their ancient past. For other scientists, the value must be less than 10^{-9} for them to be convinced that two sequences are related through evolution.

The third section in this report shows the actual amino acid–by–amino acid alignment of your input sequence and the sequence in the database that shows some similarity. Your input or "query" sequence is shown on the top and the matching sequence from the database is on the bottom. In between is a line that indicates each amino acid that is identical at that position. If the amino acids are merely similar, such as an alanine in one sequence and a valine in the other, a "+" appears in the middle line. If the amino acids at a position are not at all similar, there is just a blank area in the middle line.

There are numbers associated with both your input or "query" sequence and the matching sequence or "subject" that was found in the database. Be careful when looking at such numbers to notice whether or not the numbers correspond to bases, even if the amino acids are shown. In a tblastn result, even though the alignment shows two amino acid sequences, the numbering for the subject (the bottom row) will reflect the nucleotides in the processed mRNA after all the introns have been cut out. This is because a tblastn search asks the computer to look at all the deposited base sequences and translate those bases so as to create all possible protein sequences. Because one can start at base 1, 2, or 3 and use either side of the DNA, there are six possible amino acid sequences that can be generated from the same piece of DNA. Each of these options, whether you start at base 1, 2, or 3, is referred to as a reading frame. However, while there are many possible sequences to consider along a particular stretch of DNA, only one is likely to be used, if at all, to code for a protein in higher organisms. It is curious that in some viruses, evolved to make economic use of their genetic information, more than one reading frame is used to code for proteins along the same stretch of nucleic acid.

Stop and Consider

How does the National Center for Biotechnology Information's Website (*www.ncbi.nlm.nih.gov*) help scientists around the world share information and further their research?

Proteins May Share Sequence and/or Structural Similarity

Before we look at an amino acid–by–amino acid comparison of KcsA with channels in other organisms, it is important to understand the ways that two different proteins can be similar.

Amino acid similarities are relatively easy to score, but underlying protein folding similarities called **secondary structures** are harder to pick out. Websites accessed through the **ExPASy**, or **Ex**pert **P**rotein **A**nalysis **Sy**stem Proteomics Server, at *http://www.expasy.org* are dedicated to picking out other features such as structural similarities based upon the DNA sequence. For example, a subset of amino acids within a string of amino acids may create a particular architecture such as an alpha helix or a beta sheet. You have already seen several alpha helices (plural of helix) in the cartoon for the KcsA potassium channel. Each of the four protein chains has three alpha helices: two long helices near each end and a short helix that sits behind the selectivity pore in the middle.

The **alpha helix** (Figure 6.5) is made up of a sequence of amino acids that can easily wrap around itself as if to make a cylinder. Each and every amino acid within the cylinder is connected to an amino acid four amino acids ahead and an amino acid four amino acids back. These loose connections are made through hydrogen bonds. Hydrogen bonds are weak bonds that in this case are made when the **amino hydrogen** of each amino acid is shared with the carbonyl oxygen of another amino acid four amino acids away. You can see from the figure that the backbone of the alpha helix spirals to create a cylinder. If we turn the cylinder on end, we see that all of the side chains stick out away from the center. Often, but not

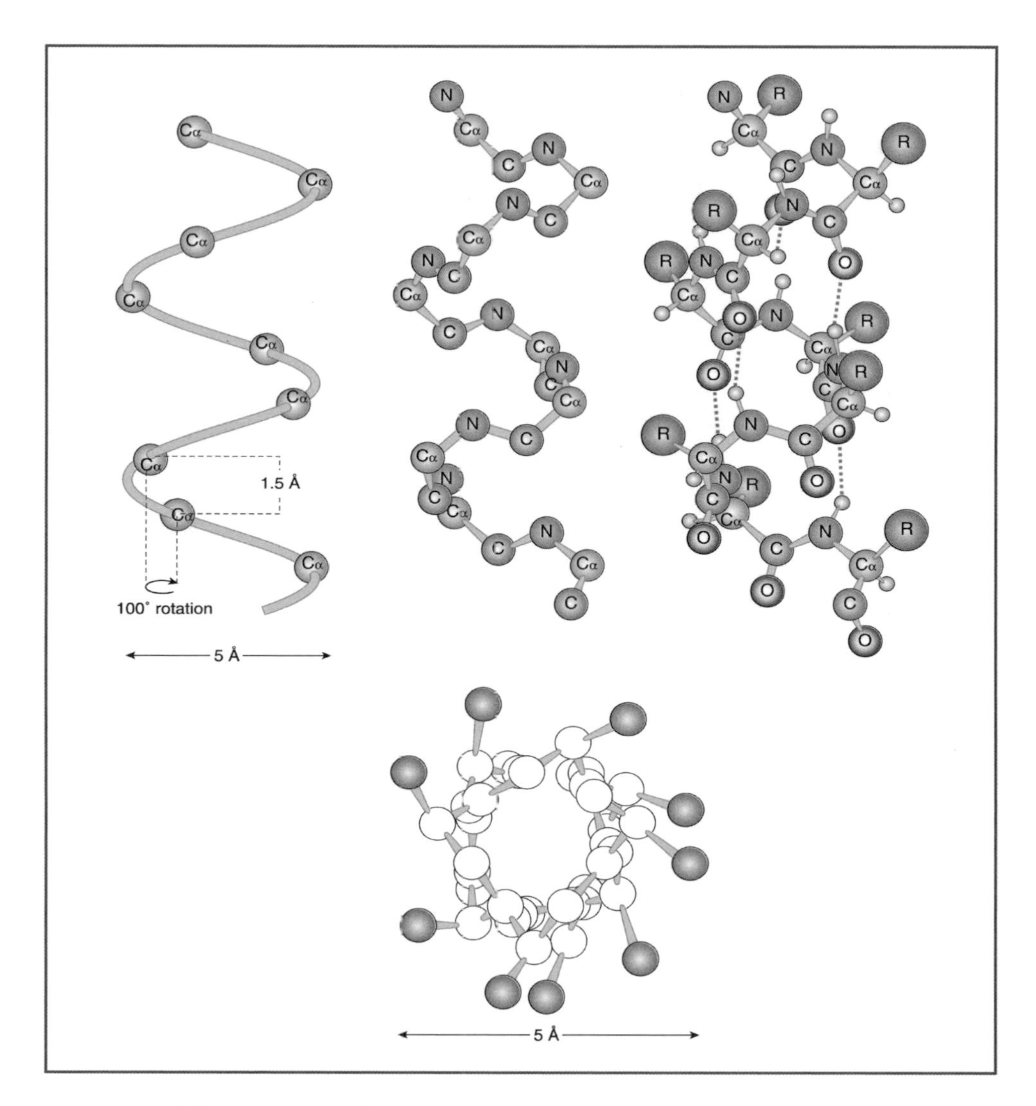

Figure 6.5 The structure of the alpha helix is illustrated here. The backbone of an alpha helix is held in place by hydrogen bonds shared with amino acids four residues away (top). On the left is shown only the middle carbon within each amino acid; the entire amino acid backbone is shown in the middle, and the hydrogen bonds are indicated in the far right structure with dotted lines. Three and a half amino acids are required for each complete turn of the helix. The side chains stick out away from the cylindrical backbone (bottom).

always, one side of the cylinder is hydrophilic while the other side is hydrophobic and greasy. For example, the greasy side might contact a membrane which is similarly greasy while the hydrophilic side might contact the cell cytoplasm, the semifluid inside of the cell. The cytoplasm itself is full of polar water molecules that are most content (energetically stable) when they can form hydrogen bonds (share hydrogens) not only among themselves, but with whatever is in the neighborhood. Since every three and a half amino acids encircle the cylinder one time, the primary sequence of amino acids will have to alternate between hydrophilic and hydrophobic amino acids in a peculiar way to create this effect. With sophisticated computer programs or algorithms programmed to find such sequences, we can determine the sequences of amino acids that are likely to form an alpha helix within a particular protein, even when the structure of that protein is not known. Note that because there are many possible primary sequences of amino acids that can generate this effect, the primary sequences of two protein chains may be very different while the structures these proteins assume may be quite similar. Because protein structure defines its function, it is good to keep in mind that a structural match without a sequence match may indicate portions of two very different proteins that carry out the same function.

Another common secondary structure is the **beta sheet** (Figure 6.6). Here the primary amino acid sequence curves back upon itself to create a sheet of protein that looks like corrugated cardboard. The side chains stick out at the peak of the fold on either side so that one side, for example, might be greasy while the other side might be hydrophilic. In order to achieve this effect, the primary sequence must have a peculiar pattern, which computer algorithms are able to discern.

Functional domains within proteins are structures that are formed by a portion of a protein that carry out the same function and are usually the same shape from one protein to the next. In the next

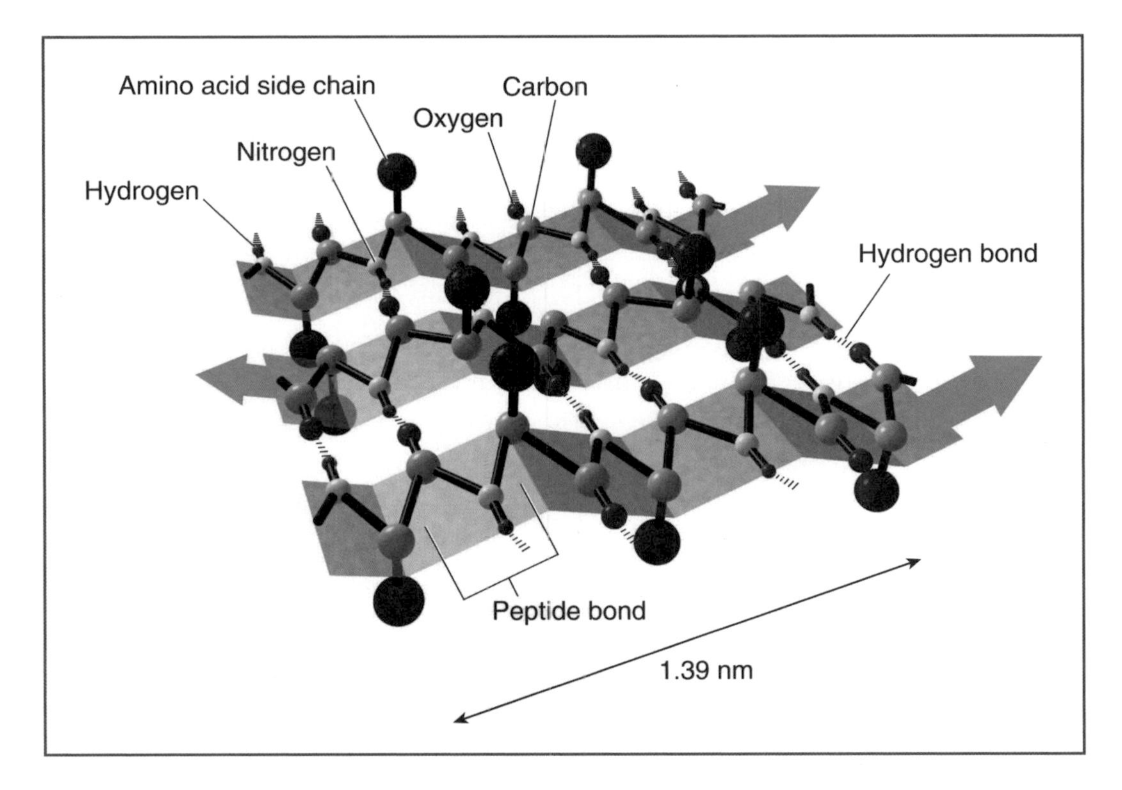

Figure 6.6 A single polypeptide folded back on itself can create a beta sheet, shown here. The side chains stick out in both directions away from the corrugated surface and the rows are stabilized with hydrogen bonds.

chapter, we will examine a functional domain that binds the cell's energy molecule ATP. Often functional domains are made up of a particular combination of alpha helices and/or beta sheets.

We have already seen that the amino acid **sequence motif** GYGD, along with other sequence features, can predict that a particular sequence will code for an ion channel. Through practice, scientists have honed their ability to predict a great variety of structural domains and sequence motifs that indicate something about the structure and function of a protein. Proteins that have similar structures are said to form a **family** of proteins. Scientists can also use sequence motifs to predict the beginning and end of a gene, the sequence that codes for one or more mRNAs and that in turn will determine proteins. Therefore, the DNA sequences deposited in GenBank can yield a great deal of information about protein machines and the genes that code for them.

The Gene Page

A quick examination of the sequences you have "pulled up" after "blasting" the KcsA potassium channel includes ion channels from a variety of organisms, including humans! Some of these sequences will have boxed letters at the end of the row. For example, the letter "G" provides a link to the Entrez Gene page summary for that particular channel. The Gene page summary will indicate other aliases, or names, for this particular gene and among other things will give the precise chromosomal location of this gene. For example, the location for the human KCNQ2 potassium channel gene that is similar to the KcsA bacterial channel is "20q13.3." This chromosomal location says that this gene is located on the long arm of chromosome 20. Because each chromosome has a slight indentation at the centromere, the place where the identical chromosomal copies attach to one another after replication, the chromosome will likely be longer on one side of this indentation that on the other. The long arm is referred to as the "q" arm, while the short arm is referred to as the "p" arm. The other numbers refer to first an arbitrary region and, secondly, light and dark bands within that region that are visible through staining when the chromosomes are tightly wrapped just before the cell divides.

The Gene page can be adjusted through a pull-down menu to show a very complete graphical display and list of important information about a particular gene. The top of the page uses a cartoon to indicate the possible processed transcripts made by this gene. Even if the gene codes for only one mRNA initially, more than one species of mRNA may result through the use of **alternative splice**

sites. Thus, while exons 1, 2, 3, 4, and 5 may be present in one transcript, only exons 1, 4, and 5 may be present in another because all of the intervening sequence was cut out of that mRNA and the rest spliced back together before it left the nucleus. As you might have guessed, the use of **alternative splicing** is not haphazard. Rather, the mRNA splicing is highly regulated. Another way to produce a different mRNA or mRNA variant is to start the mRNA at a different position, for example, inside of the normal site.

One can see the intron and exon structure of each transcript variant for a particular gene by examining the series of boxes (exons) and lines (introns) that indicate the expanse of each transcript. Blue portions of boxes show the beginning and ending portions of the mRNA that will not code for any amino acids. These portions are referred to as the **5'** (beginning) and **3'** (end) **UTRs** or **untranslated regions.** Databases define the gene as the expanse of DNA sequence that is initially copied into mRNA, including the introns, exons, and UTRs.

Because so much variability is possible based upon the DNA sequence of one gene, the Gene page helps us to keep all of this straight by providing links to reference sequences that correspond to each variant mRNA and the corresponding proteins that are made. These reference sequences are labeled under the heading RefSeq or with unique identifiers that often, but not always, begin with the letters NM_____ or NP______. These represent unique transcripts and proteins that are made when particular exons and UTRs are incorporated into the mRNA.

The Entrez Gene page also uses a cartoon to show the relative location and orientation of this particular gene on the chromosome with its nearest neighbors. A "**summary**" section is also included on this Entrez Gene page to give a quick overview of the function of the protein made by this gene. The summary for the KCNQ2 gene says that another gene, KCNQ3, is required for these polypeptides to form a potassium channel together that is called an M channel, and that a drug called retigabine can inhibit the M channel. The Gene page has many other features as well. Among them is a bibliographical section with live links to free PUBMED summaries of articles and further links to the complete journal article about that gene. In some cases, the complete article can be downloaded onto your computer for free; in other cases, one must pay a small fee to do this.

The Entrez Gene page also provides links to another service of the NCBI, the Online Mendelian Inheritance in Man (OMIM) page. If mutations in your gene are associated with a particular disease or disorder, such links will display the letters MIM followed by a number. For example the link MIM:606457 from the Gene page for the human potassium channel gene KCNQ2 provides a link to the OMIM database entry #606457 entitled "Myokymia with Neonatal Epilepsy."

By following the exercise about KcsA and KCNQ2 in the Appendix, you will learn about the work of Dr. K. Dedek and his colleagues who found that arginine 207 was changed to a tryptophan by a mutation in the DNA code for the KCNQ2 gene.[4] While each OMIM entry might have a slightly different content, a regular part of most files is a section called CLINICAL FEATURES. Here you will learn about specific mutations that cause disease. Most entries will contain references that can be accessed there or at the bottom of the page in a REFERENCES section with live links to PUBMED summaries of that research. In this way, you can learn about the normal function of that particular protein and perhaps something about the way that that mutation has changed the structure of that protein so that it no longer functions properly. For example, Dr. Dedek's work showed that the arginine mutation that caused convulsions in three generations of a family is part of a section of the KCNQ2 potassium channel that senses a build-up of charge in the membrane. He knew this because Roderick MacKinnon and members of his laboratory who first described a bacterial potassium channel's structure in 1998 also determined the structure of another channel in 2003 that has floppy paddles that are pushed away when the cell lets in positive charges near the membrane.[5] Because the paddles are connected to the channel's gate, this movement opens the channel gate. The floppy paddles have a series of positively charged amino acids along one side that sense the accumulation of positive charges within the cell when a nerve innervates a muscle and causes charges to build up inside the muscle cell. Because the KCNQ2 channel was missing a charge, the gate was less sensitive and the potassium channel opened more slowly. Evidently this bad timing causes muscle spasms or convulsions.

CONNECTIONS

In this chapter we have come full circle from bacteria to humans in investigating a protein whose presence was demonstrated long before we got a glimpse of its universal appeal. We learned how to use the NCBI server to access information about a particular protein. As an example, we used the bacterial potassium channel KcsA. Because of our tax dollars at work and the effort of countless scientists throughout the world, we are able to access an invisible world of information available to anyone at the click of a mouse. A protein file provides the species of origin for that protein, its subunit structure, and special features encoded in its primary amino acid sequence. For the KcsA potassium channel protein, such special features include the pore helices and the selectivity filter. Bibliographic references within the protein file allow one to link to summaries of the original research about that protein. For example, we accessed a summary of Dr. MacKinnon's Nobel Prize–winning work on the KcsA structure. From the protein file, one may access the amino acid sequence of this polypeptide in FASTA format. That sequence can then be used as bait to find all other similar sequences in a BLAST search. A search using the KcsA protein sequence highlighted a human voltage-sensing potassium channel protein KCNQ2, one that has paddles that can both sense a charge build-up and rip open the channel's door. From the BLAST results, one may link to a Gene file that provides specific information about the gene specified in the BLAST result. For example, KCNQ2 makes a variety of mRNAs. Sometimes within a Gene file, there is a link to an Online Mendelian Inheritance in Man (OMIM) file that describes an inherited disorder associated with mutations in that particular gene. Myokymia, a condition in which individuals experience involuntary small undulating muscle contractions under the skin of the hands and feet, and neonatal convulsions are caused by a change in the DNA code that results in the incorporation of a tryptophan in place of an arginine in amino acid 207 of the KCNQ2 protein chain. This slight change makes all the difference in the world to the children and families afflicted. Amazingly a positively charged amino acid side chain is replaced with one with no charge, and this happens within a critical part of the channel. This changes the ability of the floppy paddles in the channel formed by KCNQ2 and another protein chain called KCNQ3 to pull the channel door open.

FOR MORE INFORMATION

For more information about the concepts discussed in this chapter, search the Web for the following keywords:

KcsA, GYGD, Evolutionary tree, Functional domains

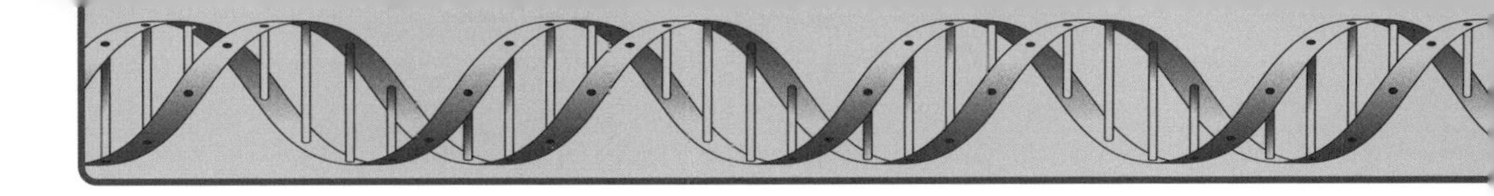

7

More About Proteins: How to Find Similar Proteins

THE ATP SYNTHASE WORKS LIKE A CAR'S DISTRIBUTOR

A colleague of mine made his living as a woodworking artisan before his attention to detail made that life untenable. He now delights our students with his singular take on mathematics. I introduced him to the general structure of the **ATP synthase** (synthase is biochemists' shorthand for a protein that makes something) protein and then asked him to view an online animation of its operation. He was duly impressed:

> This is astounding! Why has man chosen to recreate the mechanical precision of Mother Nature? The device [the ATP synthase] very much resembles the distributor of an automotive engine in both its structure and operation. . . . The distributor's rotor is driven in a circle by the rotation of the engine. As the rotor turns, it touches each of the contacts in the cap, thereby

> sending power to each cylinder [a distributor directs the sequential firing of each cylinder by farming out electrical power to each cylinder in turn]. It is remarkable that humans have invented a mechanical device that so resembles an organic device favored by evolution.[1]

In this chapter you will not only learn more about this distributor protein, but you will see that we are not very different from chimps or even mice or rats when it comes to these critical housekeeping proteins that keep our cells going.

The amino acid sequences of many cellular machines that carry out basic processes common to all forms of life have been maintained with only slight variations from their initial appearance in single-celled organisms 3 billion years ago. The ATP synthase is one of those essential machines. The recent sequencing of one of each type of chromosome in over a thousand organisms from the lowly bacterium to humans has made this patently clear. Many of the protein machines that are essential to the life of a bacterium have been preserved to carry out the same functions in more complicated life forms such as higher plants and mammals like us. While the essence of each machine is maintained, the slight adjustments accommodate new interacting partners that allow them to function in different organisms. Scientists have been pleased to find that old evolutionary trees based on comparison of the outward appearance and internal structures of various plants and animals have been borne out by considering the progressive alterations in their coding sequences.

In this chapter, you will learn more about how to use the NCBI Website, in this case to ask questions about this protein that rotates like a car's distributor. Note that similar kinds of questions can be answered about other proteins of interest by following the same sequence of steps. First, you will learn how to answer the question, "How widespread is this protein?" and view a molecular model of the ATP binding region. Then, you will learn how to answer the question, "How similar are the human and chimp ATP synthase proteins?" and examine the alignment of a human and chimp synthase subunit. Finally, you will see how to compare the chromosomal neighborhood of this gene in humans, mice, and rats and access other information that addresses questions about this or any other particular gene product.

Making ATP is serious business for most, if not all, cells. **ATP** (***a**denosine **t**ri-**p**hosphate*) is an adenosine molecule with three phosphate groups attached to it. The energy released when the outermost phosphate is broken off is used to power many reactions in the cell. Therefore, the cell's batteries are recharged by reattaching that outermost phosphate. The protein machine that does this is ATP synthase. The shape and function of the ATP synthase, that recycles energy molecules of the cell by snapping together ADP and inorganic phosphate to make ATP, is now

ATP Synthase

The 1997 Nobel Prize in chemistry was awarded to Dr. Paul D. Boyer of the United States, Dr. John E. Walker of the United Kingdom, and Dr. Jens C. Skou of Denmark. Learn about their work that demonstrated the mechanisms of the ATP synthase by going to *http://nobelprize.org/chemistry/educational/poster/1997/boyer-walker.html*. Then view the ATP synthase in action at the Website of Professor Wolfgang Junge at the University of Osnabrück in Germany at *http://www.biologie.uni-osnabrueck.de/Biophysik/Junge/overheads.html*. These movies run using a free, downloadable QuickTime™ plug-in. The "F_0F_1-ATPSynthase (animation)" clearly shows in three dimensions the working of the entire ATP synthase, including the membrane-embedded F_0 portion and the three F_1 bulbs each containing an alpha and beta domain. To break apart the animation, examine his movies "F_1-ATPase (animation)" and "F_0 (animation)." The "Rotation of F_0 in F_0F_1 (movie)" demonstrates the magnified real-time turning of the F_0 portion as indicated by a jerking, rotating actin fiber that was attached to one of the F_0 subunits and made visible by a fluorescent tag.

known through the Nobel Prize–winning work of Paul D. Boyer,[2] John E. Walker, and Jens C. Skou. Indeed, this protein has a mechanical essence. Like many proteins, it is made up of several long molecules, or polypeptides, each made up of a long string of amino acids. Scientists often distinguish these polypeptides from one another by assigning the subunits names that are letters of the Greek alphabet. Figure 1.2, presented earlier, illustrates the contributions of each of these chains to the whole ATP synthase. The membrane-embedded or **F_0** portion has multiple copies of chain c along with a few other subunits nearby. Attached to the surface of the F_0 portion is a spindle made up of the gamma and epsilon subunits. As the spindle turns, it differentially distorts three balloon-like subunits on top.

This machine's function is to recharge the cell's chemical energy battery (ATP) by connecting a negatively charged **phosphate group** (**P_i**) to an already negatively charged molecule called **ADP** (**adenosine di-phosphate**). Not only are these molecules repulsed by each other, but the thought of joining them so that their electrons are rearranged into a stable configuration is daunting.

Not so for the ATP synthase. Its job is precisely that described above, to make ATP from ADP and P_i. How does it do this? The top part, or **F_1**, section of the ATP synthase sticks out away from the membrane. It is made up of three functional sections that look like three flexible balloons squashed together. Each ballon is made up of two protein subunits. The raw materials for this synthesis enter each balloon from the side. As the spindle turns, it deforms each of the three balloons in turn. A single site goes from a welcoming "loose" configuration that receives the ADP and phosphate to a "tight" configuration that squeezes the unlikely bedfellows together inside the balloon to recharge the ADP. As the spindle continues around, the newly formed ATP is free to float away from the "open" configuration and provide room for a new arrival of ADP and P_i (Figure 7.1).

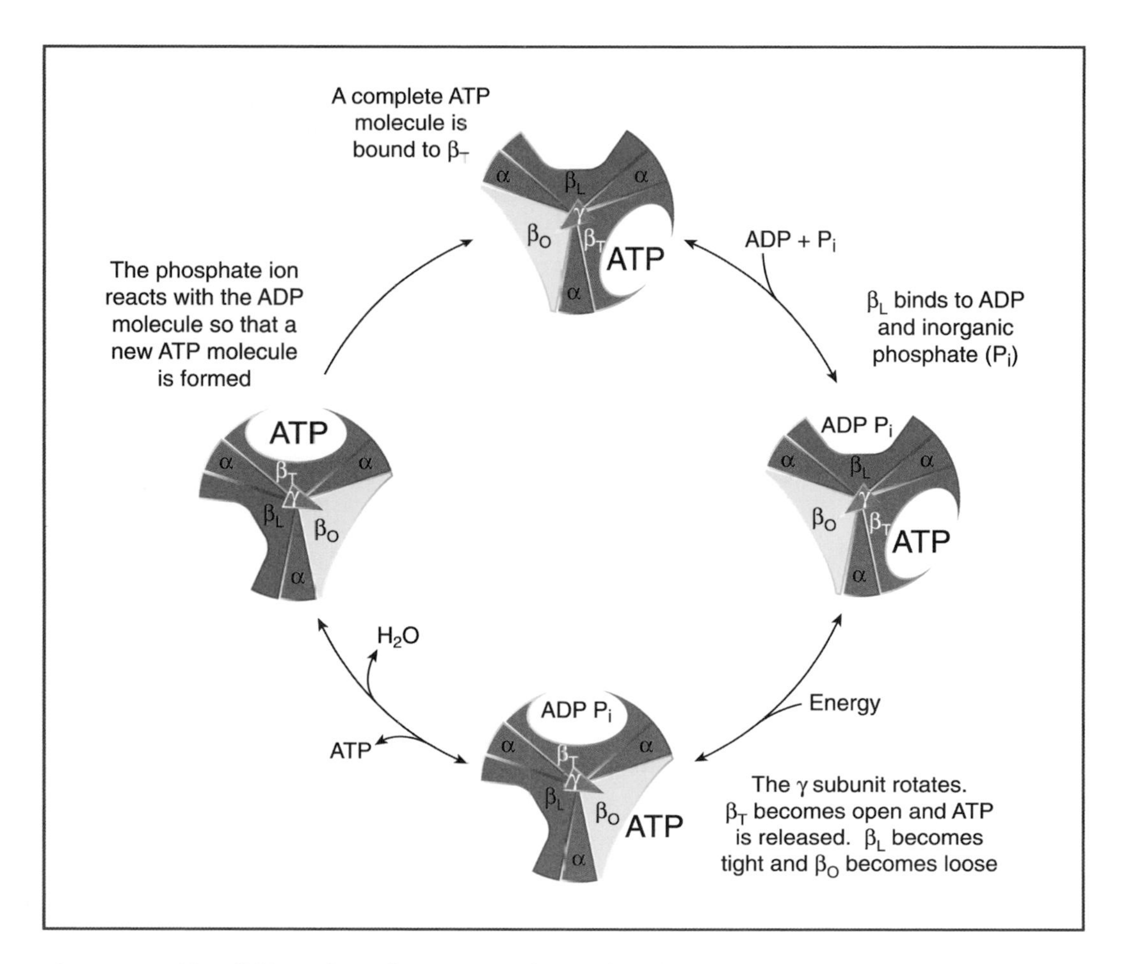

Figure 7.1 The ATP synthase is a universal machine that combines ADP with inorganic phosphate to make ATP. This view from the bottom of the F_1 portion shows the protein subunits indicated by the Greek letters α, β, and γ. As hydrogen ions pass from one side of the inner mitochondrial membrane to the other, they cause the F_0 portion of this protein machine to turn. This makes the central shaft turn, deforming the alpha-beta bulbs of the F_1 portion that squeeze ADP and phosphate together (T = "tight," L = "loose," O = "open"). *Source: R. MacKinnon, "Nobel Lecture. Potassium Channels and the Atomic Basis of Selective Ion Conduction,"* Bioscience Reports *24 (2004): 75–100.*

All of this movement is fueled by a 10- to 100-fold build-up of hydrogen ions. Living things chemically break down food to create this imbalance in a large membrane-bound compartment within the **mitochondrion.** (A similar imbalance in the **chloroplasts** of plant cells is powered by light.) We pack high-energy electrons from burgers and fries onto chemical carriers called NADH and $FADH_2$. Notice the hydrogen, H, in these chemicals. These carriers hand the electrons off to proteins in the convoluted inner membrane of our mitochondria. In response to this gift, the proteins pump hydrogen ions into the membrane-bound compartment and the hydrogen ion concentration builds up in the enclosed space. The ATP synthase is fortuitously positioned on the membrane. When the crowded hydrogen ions chance to enter the bottom portion of the ATP synthase (called F_0) embedded in the membrane, the ratchet advances, just as a car's engine turns over. (In a Model T Ford, the motor was literally "turned over" for the first time with a hand crank.) The turning of the F_0 rachet advances the asymmetrical central rod of the ATP synthase, just as turning the engine makes the distributor's rotor rotate. The rotation of the central rod inside the ATP synthase in turn makes each of the three bulbs change its conformation so as to continually recycle ADP + P_i into ATP. Similarly, the distributor rotor sends voltage to each contact as it rotates within the distributor cap. (This electricity will sequentially fire the cylinders that push the pistons that keep the motor turning.) Because ATP is the common currency of energy exchange for organisms as diverse as bacteria and humans, one would expect to find the code for this machine in the genomes of a variety of organisms. As you will see below, this machine is indeed found in organisms as simple as bacteria and as complex as humans.

Stop and Consider

Why is ATP synthase so important? What do you think would happen if a person had a mutation that caused ATP synthase to work incorrectly, or that person's body did not make enough ATP synthase?

HOW WIDESPREAD IS THE ATP SYNTHASE?

There are several ways to determine how common your protein of interest is in all living things. But one of the most effective ways to access many avenues of information about your particular protein is to use the "All Databases" option when searching the NCBI server. At *http://www.ncbi.nlm.nih.gov/*, you can adjust the display window to read "All Databases" and then type in the name of the protein such as "ATP synthase" that you are interested in researching. By indicating the topic of your search and selecting "Go", you will bring up a variety of links to information related to that topic (Figure 7.2).

The NCBI Entrez page has graphical links to a variety of databases. Each of these databases is represented by an icon. The number of files pertaining to your protein of interest, such as the ATP synthase, is indicated by the number next to each icon. One of the icon options is for **HomoloGene**. The HomoloGene link will look through all known eukaryotic genomes for proteins similar to the one you typed in earlier. (Eukaryotes—cells that have a nucleus—include every living thing except bacteria and some other microbes called *Archaea.*) Selecting the round HomoloGene icon will bring up a page listing multiple sets of genes related to your protein of interest. For example, had you earlier typed in "ATP synthase" when searching all databases, each set would include genes that code for a particular subunit of the ATP synthase in a variety of eukaryotic organisms. The set labeled HomoloGene: 1273 lists genes that are similar to the human beta subunit of the ATP synthase (Figure 7.3).

By selecting a particular list, one can pull up a HomoloGene Discover Homologs page. A HomoloGene Discover Homologs page has several key features. It not only lists and provides links to genes in a variety of organisms that are similar to your gene of interest, but perhaps more importantly it displays a cartoon of the portions of each gene that will generate a similar structure. For example, selecting the blue line of type labeled HomoloGene: 1273

(continued on page 136)

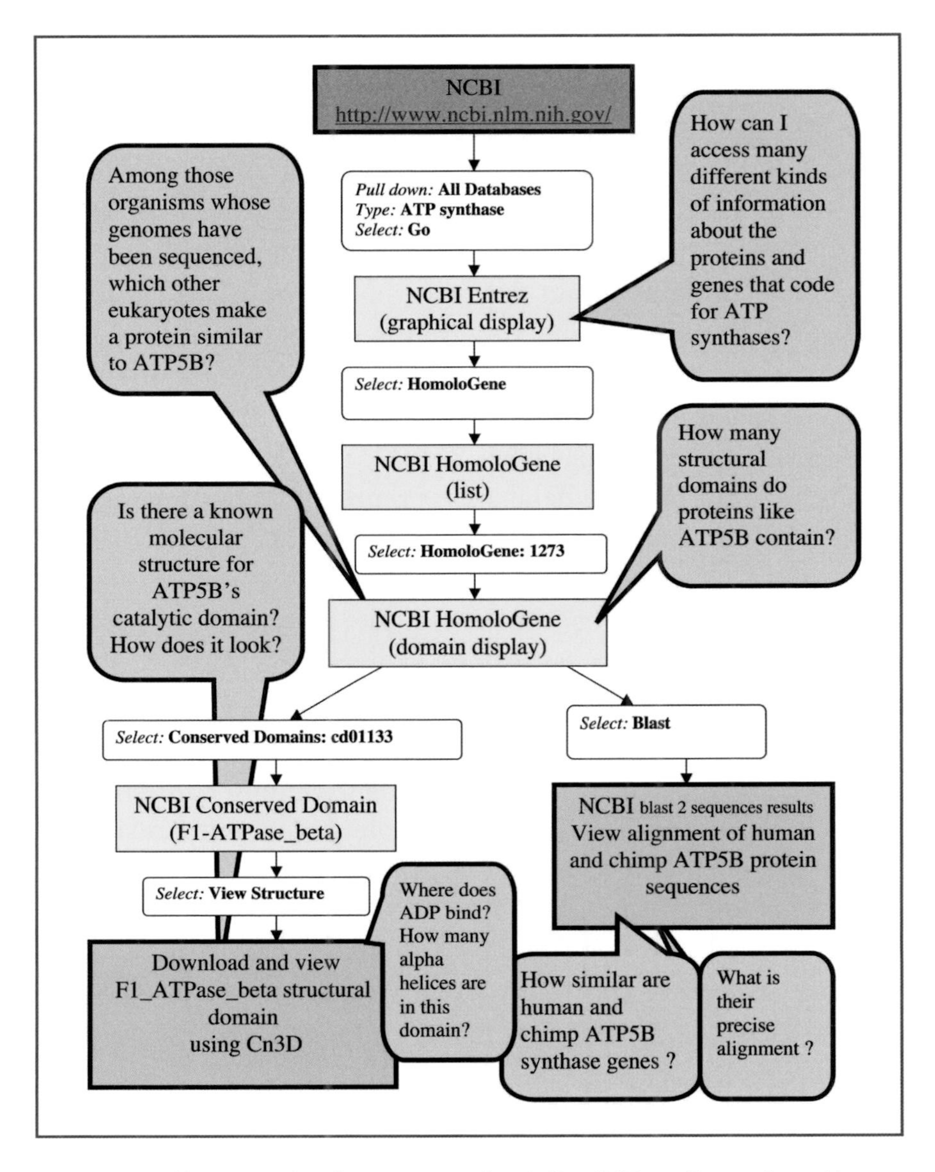

Figure 7.2 You can also learn more about the ATP synthase from the NCBI Webpage. This flow chart describes the path to compare the ATP5B subunit of the ATP synthase with similar proteins within other completed eukaryotic genomes. On the right, one is shown how to compare the human ATP5B protein sequence with that of the chimp.

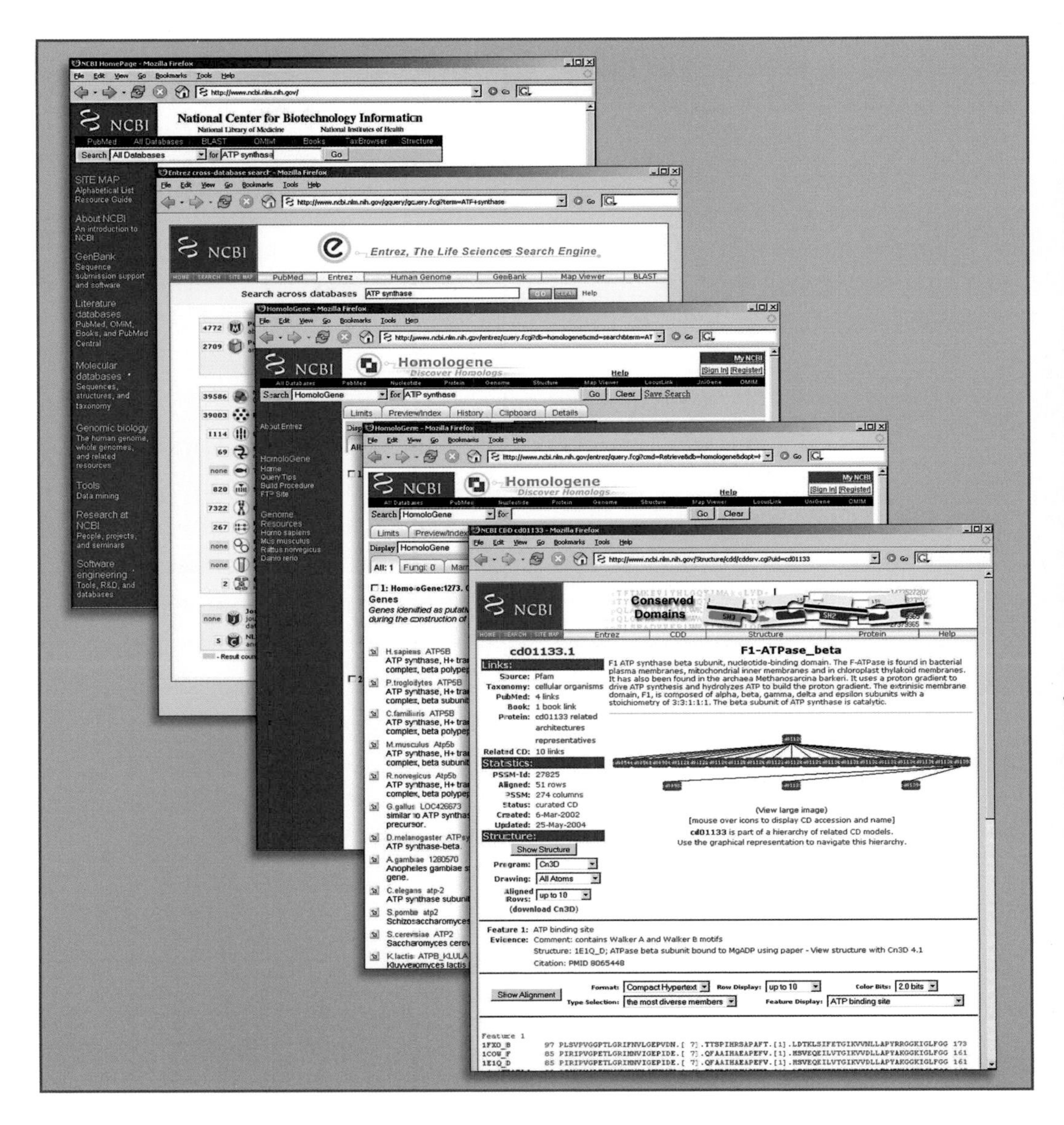

Figure 7.3 Through the NCBI Webpage, you can access the F1-ATPase_beta structure. The pages encountered in accessing the cd01133.1 F1-ATPase_beta structural domain are shown here, which correspond to the flow chart given on the left side of Figure 7.2. Selecting "View Structure" will bring up the atomic structure of this domain, which is shown in Figure 7.4 on page 139.

(continued from page 133)

would bring up a set of similar genes graphically represented to show equivalent internal structural domains. These sequences may include similar proteins from baker's yeast (*S. cerevisiae*) all the way to humans (*H. sapiens*). For example, there is a similar ATP synthase beta subunit gene in flies and mosquitoes (*D. melanogaster* and *A. gambiae*), rats (*R. norvegicus*), mice (*M. musculus*) and chimps (*P. troglodytes*). Had we looked in a microbes database, we would have found this ATP synthase subunit gene there as well.

WHAT DOES THIS ADP-BINDING DOMAIN LOOK LIKE?

As discussed above, the cartoon on the right on an NCBI HomoloGene Discover Homologs page indicates parts of the protein that code for similar structures. The structure of each of these domains has already been confirmed through their similarity to the **consensus sequence**, the most likely sequence for an entire family of proteins. The consensus sequence highlights certain amino acids that are always the same and other amino acids that may vary in a limited way in terms of their properties among similar proteins in different organisms. These position-specific rules for the sequence create a pattern that a computer can recognize. For example, two different proteins that have a similar, or **homologous**, structural domain might always have a cysteine at one particular position and a serine at another. It is likely that these invariant amino acids (scientists say "these amino acids are highly conserved") are very important in contacting a specific molecule that allows this protein to do whatever it needs to do. Therefore, consistency at this position within the protein is absolutely required. In some cases, the amino acid at a certain position might always be positively charged. Therefore, this amino acid might be arginine in one protein and lysine in another, but it must always be one or the other. Other positions within the protein might always be occupied by hydrophobic amino acids, such as leucine, isoleucine, or valine. Here, the hydrophobicity, not the exact size or configuration of the amino acid, would be important.

Many of these domain structures are made up of a combination of alpha helices, beta sheets (both discussed in Chapter 6), and loops linking them together. Because there are many sequences that can form such structures, the amino acid–by–amino acid sequence may be very different. Even though there may be no more than 30% identity among the sequences of individual proteins that contain this structural domain, the structures formed may nonetheless be quite similar because the substitutions within the sequence still allow the overall structure to be maintained. At least eight major databases are devoted to organizing and identifying structural domains and other types of patterns and there are three servers for accessing this information. By using the NCBI server, you are automatically able to access the NCBI Conserved Domain or CD database, the Pfam or *p*rotein *fam*ilies and domains database, and the SMART databases. The PfamA database is maintained by a largely English, Swedish, American, and French effort and is the largest manually generated domain or profile database. SMART stands for *S*imple *M*odular *A*rchitecture *R*esearch *T*ool and is a recent effort to catalogue domains used in the extracellular matrix (proteins secreted by animal cells), in cell signaling (proteins that form a relay team of activation between the membrane and the inside of the cell), and domains within proteins that contact DNA.

On the HomoloGene Discover Homologs page for the ATP synthase beta subunit protein group 1273, selecting each of the colored bars in turn will allow you to view information about each one of these known conserved protein domains. The fact that the shape of a particular part of a molecule has been preserved through millions, and sometimes billions, of years of evolution, generation after generation, suggests that particular configuration of amino acid sequences produces a shape that has an important function. One might select a domain on the graphic. Alternatively, one might select the domain of interest under the heading “Conserved Domain” lower down on the page. Either action will bring up a file labeled with the prefix "cd" from the NCBI's *C*onserved *D*omains Database, the prefix "pfam" for the Pfam protein families database, or the prefix "smart" for the SMART protein architecture database. The file will list the amino acid sequence of this particular domain within the sequence of individual members that have that domain and the consensus sequence for that particular structural domain. For example, selecting the middle cartoon highlights the F1-ATPase_beta domain[3] file cd01133. This action brings up individual protein sequence files aligned under the consensus sequence that describes this family of proteins.

One of the best ways to examine such domains is to look at a model of the structure. Links to such structures are often provided at the bottom of the conserved domain file under the heading "Show Structure." However, an even better view that provides structural evidence for a critical function of the protein such as binding the molecules it will act upon is sometimes provided separately as a "Feature" that includes the command "View structure." For example, one might examine a structure that shows the ATP synthase in contact with its substrate ADP by selecting such a link. In order to view these structures, you must download a free program called Cn3D. You can find the Cn3D program by clicking on the Structure on the black bar of the NCBI page. This program allows you to not only "see" molecular structures, but manipulate them by turning them over, zooming in, and even highlighting various residues that are also highlighted on a correlated alignment. The picture of the ADP-binding site within the beta subunit of the ATP synthase is shown in Figure 7.4. This is a beautiful representation of the **catalytic site** of a cow's beta subunit of the ATP synthase bound to ADP, the molecule the ATP synthase will recharge with phosphate to create ATP.

Looking again at Figure 7.4, the green residues are the amino acids that provide a snug fit for the yellow ADP. A negatively charged amino acid called glutamic acid is highlighted in yellow. This amino acid holds the attention of the positively charged magnesium ion that

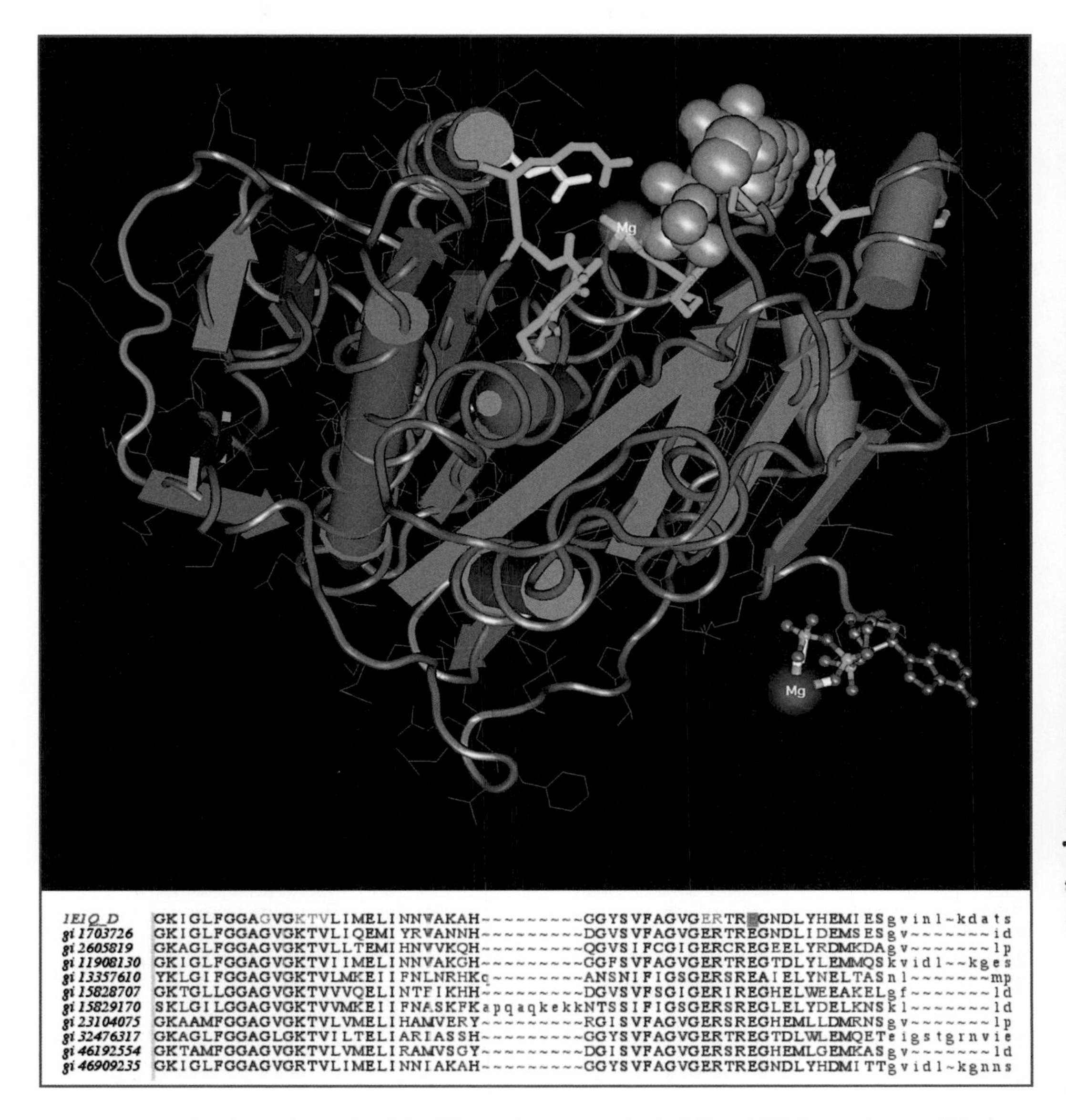

Figure 7.4 This view of the the F1-ATPase_beta domain holding ADP is made possible by the program Cn3D and is one of three identical structural domains within the ATP synthase. This conserved domain formed by the beta subunit is cd01133.1. The cell battery molecule ADP (indicated with balls) fits snugly within the active site. The sequences below highlight the conserved residues and have live links to data regarding those proteins.

itself is snuggled near negatively charged phosphates in ADP. When you bring up such a picture, you will also see a **Sequence/alignment viewer**. This tool provides all the details; whatever you highlight in the structure will be highlighted there as well, and all the residues are color-coded in both places for your convenience. You may toggle the side chains of the amino acids in the structure on and off to make them appear or disappear at will. You may change the representation of the backbone; in this rendering, the alpha helices and beta sheets are represented by crayons and flat arrows. The Cn3D message log, which is not shown in this figure, can be called up as a menu option to describe each atom you point your cursor at and compute the infinitesimally small distances between them.

HOW SIMILAR ARE THE HUMAN AND CHIMP ATP SYNTHASE BETA SUBUNITS?

From the HomoloGene Discover Homologs page for a particular group of similar proteins, one may also answer the question, "How similar is the human form of this protein to a homologous protein found in other organisms?" Homologous proteins have a similar sequence and therefore are likely to have a common ancestor.

For example, one may ask, "How similar is the human ATP synthase to the ATP synthase found in chimps?" A tool to make such comparisons is called "Alignment Scores" and is found in the middle of the HomoloGene Discover Homologs page. Two pull-down menus allow you to select the protein sequences that you want to compare. For example, you might compare the protein coded for by the human ATP synthase beta subunit gene and the homologous chimp protein.

Selecting the "BLAST" button will produce a one-to-one sequence alignment of these two proteins, for example, the human ATP synthase beta subunit amino acid sequence aligned with the comparable chimp protein sequence. The human ATP synthase beta subunit protein sequence is the "query" and the chimp protein compared against it is the "sbjct" (subject). The sequences show two strings of amino acids predicted by each gene's string of bases in the DNA. The properties of the different amino acids and their one-letter codes were discussed earlier. Suffice it to say that the ability of the ATP synthase to bind ADP depends entirely upon the integrity of the sequence of amino acids that make up this protein. It is more than interesting to note that the two sequences are almost identical. A letter between the query and the compared sequence indicates that at this position, the amino acids are identical. Of the 468 amino acids in the chimp sequence, only four are different from those of the human protein. But with a large **gap,** where there is human sequence and no chimp sequence, the score for this sequence match is only 87% **identities** (464/529 amino acids that are identical). The gap may indicate a real section present in the human protein that is not present in the chimp protein, or the chimp sequence may not be complete in this portion of its genome.

The high bit score of 2254 and the low Expect score of 0.0 both indicate that this sequence similarity is meaningful. As mentioned earlier, the Expect score reflects the likelihood that the second sequence is unrelated to the first. Therefore, an expectation of 0.0 in the comparison between the human and chimp sequences suggests that these two sequences, only slightly divergent today, likely originated in some ancient ancestor of both chimps and humans and have been conserved throughout the intervening generations of both species.

One can use the "Alignment Scores" tool to examine other suspected evolutionary relationships among similar proteins in other organisms. The rat and mouse proteins are also 529 amino acids long and 95%–96% of the amino acids are the same and, of course, are in the same order as those in our ATP synthase beta subunit. Even the ATP synthase beta subunit from the tiny roundworm *Caenorhabditis elegans* (*C. elegans*) is 81% identical within the region of 526 amino acids that can be aligned. (The worm's ATP synthase beta subunit is a bit longer than the human one so that the first 12 amino acids of the worm's 538 amino acids in this protein do not match up at all with the human protein.) However, the worm protein has 86% **positives** within this aligned region. Positives are amino acids that are only slightly different in their properties; for example, a leucine is present in the human protein where there is an isoleucine in the worm protein. Such differences will only slightly affect the function of the protein. As a sidenote, the tiny, transparent roundworm *C. elegans* is considered a model genetic organism because it has been easy to use this worm in genetic studies. Members of the "worm community" have devoted their lives to studying each of the worm's genes (previously one at a time!) and the proteins each encoded. In addition, they have completely mapped the fate of each of its approximately 1,000 cells from their origin in the fertilized egg!

Another way to compare your sequence specifically with those from another organism is to go to the "BLAST" link from the NCBI home page as discussed in the previous chapter. From the BLAST link, one may choose to blast only a subset of all the organisms whose gene and protein sequences are available. Once you have pasted in the FASTA formatted protein sequence of the human ATP synthase beta subunit ATP5B, for example, you may select the plant *Arabidopsis thaliana* to search.

Amazingly, if one examines the alignment of our ATPase beta subunit gene against a similar sequence in the weed *Arabidopsis thaliana* (you guessed it, another model genetic organism with its own devoted community of workers), the sequences again are similar. Seventy-four percent of the *Arabidopsis* amino acids are identical to those in the human protein within the aligned region. Obviously, given the similarity of these sequences, there have not been many adjustments through over a billion years of evolution since the last common ancestor of plants and humans.

HOW MANY mRNAS AND PROTEINS DOES THE ATP5B GENE MAKE?

In order to find out more about the ATP5B human gene for the beta subunit of the ATP synthase, you can go back to the NCBI homepage at *http://www.ncbi.nlm.nih.gov/*, adjust the pull-down window so that it says "Gene", and type in "ATP5B and homo sapiens" (Figure 7.5). (This will eliminate beta subunit genes that have the same name in other organisms.) Selecting "Go" will bring up the Entrez Gene summary page about the human ATP5B. This is the bare-bones view of information about ATP5B. You can access a more complete set of data regarding the human ATP5B gene by selecting the choice *ATP5B*. This will bring up the Entrez Gene "Graphics" page that gives a comprehensive and detailed look at this gene (Figure 7.6).

As discussed in the previous chapter, the Entrez Gene "Graphics" page gives a graphical representation of all of the mRNAs made by a particular gene. For example, at the top of the ATP5B page, there is a cartoon of the single type of mRNA made by this gene flanked by two large numbers that represent the nucleotides surrounding this gene. These numbers define the **ORF** or ***o*pen *r*eading *f*rame** for this gene, the beginning and end of the DNA sequence where transcription, the copying into messenger RNA, starts and ends. The red boxes represent the portion of the DNA sequence that will actually be used to code for amino acids in the finished protein. Although the initial mRNA sequence is very long, the portions of the transcript represented by thin red lines are spliced out of the final mRNA product, bringing all of the boxed regions in line together. The genes of most higher organisms encode portions of sequence that are cut out of the mRNA before the mRNA leaves the nucleus. The portions retained are called exons and the parts cut out are called introns. A blue box corresponds to DNA sequence that is present in the mRNA but does not code for any amino acids. Here, it represents

(continued on page 146)

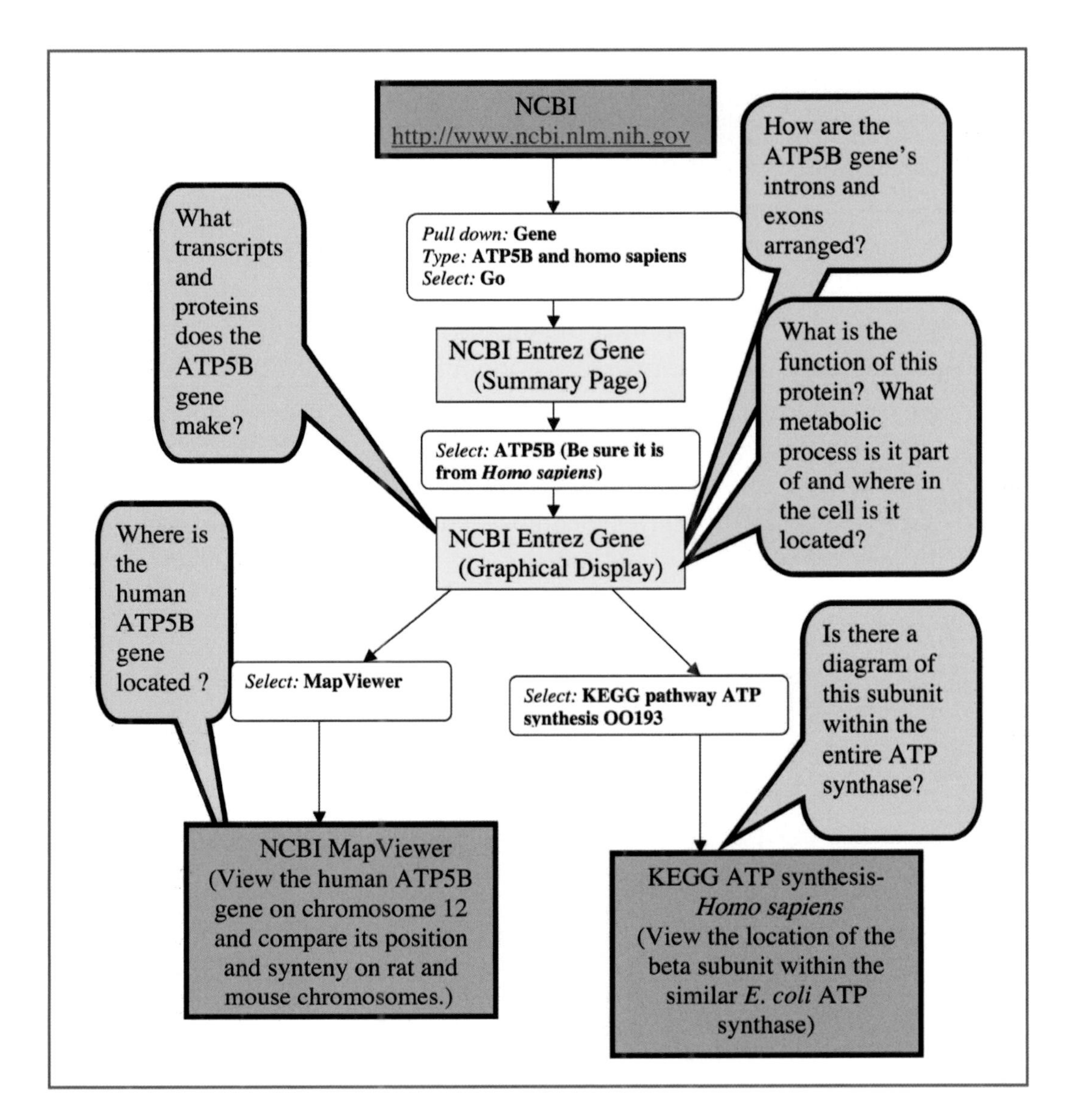

Figure 7.5 You can also use the Entrez Gene page to learn more about ATP5B from the NCBI. Follow the flow chart to compare the chromosomal neighborhood of ATP5B with that of the homologous rat and mouse chromosomes. Also use KEGG pathways to see how ATP5B interacts with other subunits within the ATP synthase and how the ATP synthase functions within a network of protein machines.

Figure 7.6 Use the pathway shown to access information about the ATP synthase beta subunit ATP5B gene. (You may also refer to Figure 7.5.) The Entrez Gene page has information about the exon and intron structure of a gene, reference mRNA and protein sequences (RefSeq), a summary of information about ATP5B, and live links to Map Viewer that show conserved regions of synteny and pathways that demonstrate interacting networks.

(continued from page 143)
the 3' untranslated region or 3'UTR. The small numbers 5' (pronounced 5-prime) and 3' indicate respectively the beginning and the end of the gene.

PRO OR CON?

Given the amazing similarity between the human and chimp sequences for the ATP5B gene and in fact an amazing similarity for the expanse of their genomes, consider the ethics of using chimpanzees and other primates in medical research. Is it right to cage these animals and subject them to medical tests, including surgery and deliberate infections, for the benefit of humans? Has the scientific community's view toward the use of chimpanzees as experimental animals changed since the human and chimp genomes were published? Where should society draw the line on such experimental research? Are some procedures acceptable but not others?

Although the ATP5B gene has several exons spread out over a large span of DNA, it codes for only one protein and in fact this is a precursor protein. It turns out that the first 47 amino acids are chopped off to create the mature protein. Many proteins have transit peptides such as this that are used to target them to a particular location within the cell. The Entrez Gene "Graphics" page also provides links to the possible mRNA and protein sequences associated with a particular gene. These sequences are often listed as reference sequences that have unique RefSeq identifiers made of letters and numbers connected by an underline. For example, the RefSeq numbers NM_001686 and NP_001677 indicate the only mRNA and protein sequences coded for by the ATP5B gene. Below this graphic of the exons and introns, the Summary may provide information about the function of a particular gene. For example, the Summary on the Entrez Gene "Graphics" page for ATP5B tells us that this gene encodes a subunit of the mitochondrial ATP synthase and summarizes information about the F_0 and F_1 portions of the ATP synthase. It states that the subunit coded for by this gene is a part of the catalytic core of the enzyme. Another section of the Entrez Gene "Graphics" page is entitled **GeneOntology**. This information is provided by the European Bioinformatics Institute, another of the genome data gathering partners, which has taken on the task of assigning molecular functions to proteins, defining the broad biological process they are a part of, the structure they are a part of, and their location within the cell. All of the entries for ATP5B relate to the function of the ATP synthase within the mitochondrion.

WHERE IS THE HUMAN ATP5B GENE LOCATED?

The NCBI Entrez Gene page also provides information about the neighborhood of genes surrounding your gene of interest on the chromosome. This link describes not only the neighborhood along the human chromosome, but also the homologous mouse and rat neighborhoods as well. Toward the middle of each NCBI Entrez Gene page for a human gene, one will find a link labeled **Map Viewer** under the heading Homology: Mouse, Rat. "Map Viewer" will bring up an alignment of the human, mouse, and rat chromosomes centered around the particular gene from whose Entrez Gene page you are linking. By manipulating the choices on the Map Viewer, you can see the location of your particular gene within the context of genes on either side of it. Not only can you view the set of genes along the human chromosome, but you can also see which genes are near your gene of choice along the mouse and rat chromosomes. The neighborhood of the ATP synthase beta subunit gene on human chromosome 12 is largely duplicated around a similar mouse gene on mouse chromosome 10. The human ATP synthase beta subunit gene ATP5B, highlighted with a red line in the center of the page, is flanked by the human genes TEBP and BAZ2A. These are both cellular "housekeeping" genes that play roles in the regulation of DNA replication and the expression of certain proteins, respectively. Similarly, the comparable mouse gene Atp5b on mouse chromosome 10 is flanked by the mouse genes Tebp and Baz2a, with a small gene for a hypothetical protein called LOC432517 between the mouse Atp5b and Baz2a genes.

Conserved segments of synteny or co-alignment of a series of genes is common among related species. Entire portions of chromosomes have remained intact during the tens of millions of years since the ancient mouse and human ancestor existed, despite the duplications, translocation to other chromosomes, and flipping of chromosomal regions. There are several links out from the Map View of a gene such as ATP5B. The hyperlink indicated by **sv** will bring up an active Sequence Viewer page that provides a bird's eye view of a particular gene's sequence. This page correlates the nucleotide sequence with the coding sequence, mRNA, and other markers and allows one to analyze this portion of the chromosome in detail.

The Map Viewer also includes a cartoon of this gene's location on a human chromosome. For example, the human ATP synthase gene's chromosomal location is 12q13.13. Translating that genetic address is not difficult. In the location name, the "12" refers to chromosome 12. The long arm of any chromosome relative to the indentation known as the centromere is the q arm while the short arm is the p arm. The rest of the numbers refer to a region and then a position within that region delineated by its black and white bands that appear when the chromosome is stained.

HOW DOES THIS SUBUNIT FIT INTO THE ATP SYNTHASE?

A section headed "Pathways" on the Entrez Gene "Graphics" page shows how your protein of interest functions within the context of a protein complex and how it functions within a network of interacting proteins. For example, the ATP5B protein is a subunit of the ATP synthase and there are two KEGG pathway entries that illustrate this. The label "KEGG" refers to the Kyoto Encyclopedia of Genes and Genomes, which has a database of molecular interaction networks, including metabolic pathways, regulatory pathways, and molecular complexes. The KEGG pathway: ATP synthesis 00193[4] link brings up the image of the bacterial ATP synthase shown in Figure 1.2. This image contains a beta subunit similar to that coded for by ATP5B. In addition there are active links from each of the subunits in boxes listed below (not shown) and the beta subunit that brought you to this image is highlighted.

A second KEGG pathway: Oxidative Phosphorylation 00190[5] is available from the ATP5B Entrez Gene "Graphics" page. This link provides drawings and additional specific information about the beta subunit within the ATP synthase. This subunit-level diagram has active links regarding the entire network of proteins that pass electrons along the membrane and pump hydrogen ions into the inner membrane space so they can wander through and power the ATP synthase. Therefore, we have come full circle in our investigation of the ATP synthase, finding it in every type of organism and using the NCBI databases and their links to learn about this powerful machine and the genes that code for it.

CONNECTIONS

Chapter 7 has introduced another amazing protein, the ATP synthase, that has the power to reenergize ADP by adding a phosphate and recharging life's molecular battery. We used the NCBI databases to find similar synthase genes across the spectrum of eukaryotes. For example, we used the "All Databases" option at NCBI to find a wealth of information about beta subunits, the portion of this protein that does the hard work of putting together ADP and phosphate to make ATP.

From the NCBI HomoloGene domain display page, you can use Cn3D to visualize a structural domain such as the ADP-binding domain that is highly conserved within beta subunits. From the NCBI HomoloGene domain display page, you may also compare the amino acid sequence of the protein coded for by a particular gene with homologous proteins from a variety of organisms. Using this tool, we compared the sequence of the human protein for the ATP synthase beta subunit with that of the chimp and displayed their amazing amino acid–by–amino acid similarity.

The Entrez Gene "Graphics" page provides a wealth of information for a particular gene. The page provides a quick view of the exon and intron structure of all of the mRNAs made by a particular gene and provides RefSeq links to their nucleic acid and protein sequences. For example, the Entrez Gene "Graphics" page for the human beta subunit gene ATP5B shows that this gene makes only one mRNA and only one protein. This page also supplies GeneOntology information about the function of this protein, the broad process it is a part of, and its location. Similarly the Entrez Gene "Graphics" page provides a Map Viewer to examine the conserved segments of synteny on mouse and rat chromosomes that correspond to the neighborhood of genes around a human gene of interest. The Entrez Gene page also provides links to "Pathways" that use graphics and active links to describe the location of a protein subunit both within the structure of a protein and within a network of interacting proteins. To illustrate such pathways for ATP5B, we linked to the Japanese KEGG pathways that showed our beta subunit in all its glory within the context not only of the ATP synthase machine, but in the context of the cohort of membrane proteins that create the hydrogen ion gradient that allows the ATP synthase to hum with activity, recycling the chemical packets that energize life itself.

FOR MORE INFORMATION

For more information about the concepts discussed in this chapter, search the Web for the following keywords:

ATP, ATP synthase, ADP, Open Reading Frame

To learn about the work of the 1997 Nobel Prize winners in chemistry, Dr. Paul D. Boyer and Dr. John E. Walker, and to learn about the mechanism of the ATP synthase, go to *http://nobelprize.org/chemistry/education/poster/1997/boyer-walker.html.*

Using Protein Explorer to Learn More About BRCA1

Go to Protein Explorer FrontDoor at *http://molvis.sdsc.edu/protexpl/frntdoor.htm#top* to use Protein Explorer to investigate the structure of the BRCA1 gene and its interaction with the BACH1 DNA helicase. You will have to first download the plug-in for a program called CHIME that allows you to visualize molecules. (Unfortunately, most newer Macintosh computers will not be able to access this information, since an old version of the operating system is required. However most PC computers are compatible with this Website.) Type in the PDB Identification code 1T29 (note that the first character is the number "1," not the letter "l") and select "Go."[1] Be sure to give the molecule time to load. You will see the BRCT domain from the BRCA1 gene with a short peptide from the BACH1 DNA helicase that it binds. This will appear as a spinning figure that has many red balls around it. To save memory, select "Toggle Spinning" to stop the revolutions and select "Hide/Show Water" to remove the red balls. The serine on the BACH1 protein is highlighted in this model by the presence of a phosphate group. Practice moving the molecule around with your mouse.

On the left of the screen, select "PE Site Map" and in the window that appears, choose the option "*Seq3D*." A new three-part window will appear at the left. The bottom window contains an interactive list of the sequences represented in the figure on the right. You will use this list to see the snug fit between the often mutated and well-conserved residues Arg1699 and Met1775 on the BRCA1 gene and the phenylalanine of the S-X-X-F motif in the BACH1 peptide. On the window on the left, locate the "F" in the peptide which corresponds to this phenylalanine. Select that F within the "B" chain sequence of the peptide portion of BACH1. When you click on it, you will notice that the side chain of the phenylalanine appears! Move the molecule around to get a good look at this residue.

Go to the top of the left-hand tripartite window and change the pull-down menu from "Element (CPK)" to "Green." Now when you select residues from the BRCT domain of the BRCA 1 gene, its atoms, except for the hydrogens, will appear as green balls. Select the residue that corresponds to Arg1699. This award-winning Website has made this easy, because as you scroll over the letters, the residue name and number appears both at the bottom of the tripartite window and in the middle window. You will see the green balls of this amino acid appear. How close are they to the phenylalanine residue in the peptide? Notice the snug fit. What would happen if Arg1699 were changed to a bulky and uncharged tryptophan?

Now change the color selection to brown and highlight residue Met-1775 and make it appear as brown balls. Which part of the S-X-X-F motif does this methionine cradle? This highly conserved residue linked to cancer also abuts the phenylalanine of the peptide.

REFERENCES

1. Protein Data Bank file 1T29 of the BRCT domain binding the S-X-X-F containing peptide is from Shiozaki, et al., *Molecular Cell* 14 (2004): 405–412.

Using BLAST

In this appendix, you will use a free service called BLAST [1] to compare the 160 amino acid sequence of the KcsA[2] polypeptide to the DNA sequences from every organism that have been deposited in the gene databases throughout the world. Because multiple entries may be the same, this database of sequences has been culled so that we will search only non-redundant or nr sequences. Follow along on Figure 6.3. This figure indicates the precise steps we will take as well as the particular questions about the KcsA protein we will answer. We will also use the "Gene" and "OMIM" pages to investigate a similar protein in humans called KCNQ2 that codes for a subunit of a potassium channel.

The first question asks, "Is any part of the bacterial KcsA K^+ channel protein coded for by the DNA sequences of any other organisms?" Before using a service called BLAST to answer this question, you will need to make a file of the amino acid sequence of the KcsA channel in "FASTA format." You may retrieve the FASTA formatted file from the Protein file P0A334 following the steps shown in Figure 6.2. Save the FASTA formatted file. After going back to the NCBI homepage at *http://www.ncbi.nlm.nih.gov/*, select the tab BLAST at the top of the page to go to *http://www.ncbi.nlm.nih.gov/BLAST/*. One of the best ways to search for genes that code for proteins that have regions of similarity to the one you have in hand is to use its protein sequence to ask the computer to look for nucleotide sequences that can be translated into a protein similar to yours. Select the choice "tblastn" or "Protein query vs. translated database" and pull up "NCBI BLAST query form." Paste the FASTA format file for the identifiers and amino acid sequence of the bacterial KcsA potassium channel into the window to "BLAST" this sequence and compare it to all of the potential amino acid sequences predicted by DNA sequences present in the nr database.

After you hit the "BLAST" button, you will see an "NCBI *formatting* BLAST" window at *http://www.ncbi.nlm.nih.gov/BLAST/Blast.cgi* that will give a description of your "query" or the sequence you inserted and an

ID number for your request. Click on the "Format!" button and wait. This page will also give an estimate of the time required to process your request, which usually takes only a minute or two.

After a short time, a page labeled "NCBI *results of* BLAST" will appear. Notice that the section labeled "Query" identifies the sequence you inserted and its length, 160 letters. There is also a description of the database that was surveyed for likely matches. As you scroll down the same page, you will see a graphic representation of the sequences that matched either all or at least some of the sequences in the bacterial potassium channel. Your original sequence is presented as a ruler, which stretches the length of 160 amino acids. Each sequence in turn has its own line that is color-coded to show the extent of the match. Notice that for many matches, the area of similarity encompasses only a portion of the entire sequence. The lines at the bottom of the page represent sequences that have the least similarity to the KcsA channel. Each of these lines is an active link to information at the bottom of the page.

As you scroll down this same page, you will see a one-line summary of each of the sequences represented by each colored line. Notice that this BLAST of the bacterial KcsA potassium channel has brought up similar sequences present in not only other bacteria but present in humans as well. The first two sequences are the identical potassium channels in *Streptomyces coelicolor* and *Streptomyces lividans*. (Remember that the KcsA channel comes from *Streptomyces lividans*. It turns out that the one in *S. coelicolor* is identical.) Several of the entries are from sequences in *Homo sapiens* or humans. The significance of each match is roughly indicated by the "E value" or "Expect value," a measure that indicates the likelihood of finding such a similar sequence by chance. Any of the E values shown here, which range from $9 \text{ X } 10^{-10}$ to $9 \text{ X } 10^{-63}$, suggest strongly that these similarities did not arise merely by chance. Rather, they suggest a possible evolutionary relationship, at least for the parts or "domains" of the channel that are shared.

Scroll down the rest of the page to see the actual one-to-one amino acid alignment of the KcsA channel sequence with the amino acids of other sequences. One can select a particular alignment by selecting either the "Bit Score" next to the first information line or the colored cartoon line above. The second and one of the two closest matches shows the original query sequence and the matching sequence from *Streptomyces lividans* aligned one on top of the other. Any amino acids that are identical are listed *between* the two lines of sequence. Therefore one can quickly glance at the aligned sequences to see the identical amino acid matches.

Note that our original sequence doesn't look the same. It now has many Xs for which there is no match in the "Subject" sequence. In order to avoid pulling out meaningless matches, the computer ignores regions where the sequence lacks complexity or would likely generate many spurious matches by chance. Note that the match or "Subject" is given in nucleotide numbers, not in the amino acid number, even though the amino acid single-letter code is used. That is because we executed a tblastn search, which allowed us to bring up the nucleotide sequence that generated a matching amino acid sequence. (Why are there three times as many nucleotides as amino acids in the sequence?)

Scroll down the page to find the match corresponding to the identifier "gi/30582924/gb/BT007043.1 Homo sapiens potassium voltage. . . ," which is between sequences in our KcsA potassium channel and sequences in a subunit of a voltage-gated potassium channel found in humans. The match-up is relatively short between our KcsA potassium channel and the human channel subunit called KCNQ2. Notice however that the signature sequence for a potassium channel, the "T+GYGD" that defines the potassium selectivity filter is here, but the TVGYGD of the bacterial channel has been changed to TIGYGD in the KCNQ2 channel. That change from a valine to an isoleucine occurred during the few billion or so intervening years of evolution, but this change is a conservative one since both valine and isoleucine are very hydrophobic. Therefore, the BLAST result indicates this similarity by showing a "+" where the two hydrophobic residues overlap.

Selecting the blue button "G" will bring up a wealth of information about this KCNQ2 gene. What are some other names or "aliases" for this human potassium channel subunit? (Answer: HGNC:6296, BFNC, EBN, etc.) On which chromosome is it located? (Answer: Chromosome 20 at position 20q13.3.) Change the display window that says "Summary" to the word "Graphics" and click on "Display" or just select *KCNQ2* to bring up a page with comprehensive and specific information about the gene KCNQ2. Take your time becoming familiar with this page. The top of the page shows in graphical form the structure of a variety of alternatively spliced messenger RNAs that this gene codes for. The red boxes denote exons and the blue boxes denote the UTRs, the ends of the mRNA that are not translated into amino acids. The series of numbers on either side of the RNA boxes indicate the variant processed mRNA nucleotide sequences labeled NM__________ and corresponding protein isoforms labeled NP__________ that correspond to these five transcripts. These are reference sequences or RefSeq mRNAs that represent unique transcripts and proteins that are made by this gene when different exons and/or UTRs are incorporated into the processed mRNA.

Which genes surround KCNQ2 on chromosome 20? (Answer: CHRNA4 and EEF1A2.) It is easy to see this along this stretch of the chromosome where the relative location of KCNQ2 is indicated with a thick red arrow. Can you see two genes that are oriented in the opposite direction? (Answer: KIAA1510 and C20.) Because both sides of the DNA can serve as a template for making mRNA, genes may be oriented in either direction, going in one direction if one strand is used as a template and going in the opposite direction if the other strand is used as a template. It is reassuring to note that mRNA sequences are always given as the 5' to 3' sequence and that this orientation always corresponds to the N-terminus to C-terminus orientation of the protein.

The Summary section tells us that KCNQ2 proteins require another protein called KCNQ3 to form a type of potassium channel that is called an M channel. What anti-convulsant drug inhibits the M channel? (Answer: retigabine.) The summary and references in the extensive Bibliography section note that defects in this channel cause neonatal convulsions. You can click on PubMed links to the right of each bibliographic entry to access a complete summary of that published work. To find out even more about this disease, scroll down the page to a section titled Phenotypes. Select MIM:#606437 to link to the Online Mendelian Inheritance in Man or OMIM database entry #606437 entitled "Myokymia with Neonatal Epilepsy."

The CLINICAL FEATURES section within OMIM reveals that a single amino acid substitution in KCNQ2 can create an array of convulsive symptoms including epilepsy in newborns. Which amino acid was altered by a mutation in the KCNQ2 gene? (Answer: The entry about the work by Dr. K. Dedek and his colleagues tells us that arginine 207 was changed to a tryptophan. In mutation shorthand, this change is called R207W.)[3] That a single amino acid change can create a persistant diseased state invites the question, "What are the properties of these particular amino acids? How might that particular substitution in each of the KCNQ2 subunits that make up this channel affect its precision performance?" (Answer: The arginine side chain has a positive charge while the tryptophan is a very hydrophobic. It is possible that the charged R side chains contributed by multiple subunits have an important function in that part of the channel protein.)

In order to find out more about this particular mutation in KCNQ2, scroll down to the REFERENCES section and select the reference to Dr. Dedek's work labelled PubMed ID: *11572947*. That summary contains a link to a free, full-text article which discusses the location of the R207W change and its unfortunate effect on three generations within one family. This arginine is one of a series of positively charged residues located on a portion of this channel that is entirely missing in the bacterial channel KcsA. This portion is known to sense the build-up of charges that eventually cause this channel to open so that a signal can travel along the nerve cell to the muscle. As we have seen from Dr. MacKinnon's most recent work, this region is part of a charge-sensing paddle that is connected to the door of the channel.[4] If the paddle is defective, if it is missing one of its series of positive charges, it raises more slowly when told to do so. Therefore, the door to the channel remains shut too long, delaying potassium flow.

The quick responses of such a channel to changes in charge build-up regulate our muscular responses. Therefore, it is compelling that a mutation in a potassium channel, and particularly in the voltage sensor of that channel, would cause this precision machine to malfunction and visit seizures upon someone whose potassium ion channel machines weren't working properly.

REFERENCES

1. S. Altschul, et al. "Basic Local Alignment Search Tool," *Journal of Molecular Biology* 215 (1990): 403–410.
2. Declan Doyle, et al. "The Structure of the Potassium Channel: Molecular Basis of K^+ Conduction and Selectivity," *Science* 280 (1998): 69–77.
3. Youxing Jiang, et al. "X-ray Structure of a Voltage-dependent K+ Channel," *Nature* 423 (2003): 33-41.
4. K. Dedek, et al. "Myokymia and Neonatal Epilepsy Caused by a Mutation in the Voltage Sensor of the KCNQ2 K^+ Channel," *Proceedings of the National Academy of Sciences USA* 98 (2001): 12272–12277.

c. 8000 B.C. Crops and livestock domesticated

c. 4000-2000 B.C. Yeast used to leaven bread and make beer and wine

c. 500 B.C. Moldy soybean curds used to treat boils (first antibiotic)

c. 100 A.D. Powdered chrysanthemums used as first insecticide

1590 Microscope invented

1600 Beginning of the Industrial Revolution in Europe

1663 Cells discovered

1675 Anton von Leeuwenhoek discovers bacteria

1797 Edward Jenner inoculates child to protect him from smallpox

1857 Louis Pasteur proposes microbe theory for fermentation

1859 Charles Darwin published the theory of evolution through natural selection

1865 Gregor Mendel published the results of his studies on heredity in peas

1890 Walther Fleming discovers chromosomes

1914 First use of bacteria to treat sewage

1919 Term *biotechnology* was coined by Karl Ereky, a Hungarian engineer

1922 First person injected with insulin, obtained from a cow

1928 Alexander Fleming discovers penicillin

1933 Hybrid corn commercialized

1944 Oswald Avery, Colin MacLeod, and Maclyn McCarty prove that DNA carries genetic information

1953 James Watson and Francis Crick publish paper describing the structure of DNA

1961 *Bacillus thuringiensis* registered as first biopesticide

1966 Marshall Warren Nirenberg, Har Gobind Korhana, and Robert Holley, figure out the genetic code

1973 Herbert Boyer and Stanley Cohen construct first recombinant DNA molecule

1975 First monoclonal antibodies produced

1975 Asilomar Conference held; participants urge U.S. government to develop guidelines for work with recombinant DNA

1977 Human gene expressed in bacteria

1977 Method developed for rapid sequencing of long stretches of DNA

1978 Recombinant human insulin produced

1980 U.S. Supreme Court allows the Chakrabarty patent for a bacterium able to break down oil because it contains two different plamsids

1980 Stanley Cohen and Herbert Boyer awarded first patent for cloning a gene; Paul Berg, Walter Gilbert, and Frederick Sanger awarded Nobel Prize in chemistry for the creation of the first recombinant molecule

1981 First transgenic animals (mice) produced

1982 Human insulin, first recombinant biotech drug, approved by the FDA

1983 Human immunodeficiency virus, the cause of AIDS, is identified by U.S. and French scientists

1983 Idea for PCR conceived by Kary Mullis, an American molecular biologist

1984 First DNA based method for genetic fingerprinting developed by Alec Jeffreys

1985 First field testing of transgenic plants resistant to insects, bacteria and viruses

1985 Recombinant human growth hormone approved by the FDA

1985 Scientists discovered that some patients who had received human growth disorder hormone from cadavers had died of a rare brain disorder

1986 First recombinant cancer drug approved, interferon

1987 The first field test of a recombinant bacterium, Frostban, engineered to inhibit ice formation

1988 Human Genome Project funded by Congress

1990 Recombinant enzyme for making cheese introduced, becoming the first recombinant product in the U.S. food supply

1990 First human gene therapy performed, in an effort to treat a child with an immune disorder

1990 Insect-resistant Bt corn approved

1994 First gene for susceptibility to breast cancer discovered

1994 First recombinant food (FlavrSavr tomatoes) approved by FDA

1994 Recombinant bovine growth hormone (bovine somatotropin, BST)

1997 Weed killer–resistant soybeans and insect-resistant cotton commercialized

1997 Dolly the sheep, the first animal cloned from an adult cell, is born

1998 Rough draft of human gene map produced, placing 30,000 genes

1999 Jesse Gelsinger, a participant in a gene therapy trial for an inherited enzyme defect, dies as a result of the treatment

2000 First report of gene therapy "cures" for an inherited immune system defect. A few months later, several of the treated children developed a blood cancer

2002 Draft of human genome sequence completed

2003 First endangered species cloned (the banteng, a wild ox of Southeast Asia)

2003 Dolly, the cloned sheep, develops a serious chronic lung disease and is euthanized

2003 Japanese scientists develop a genetically engineered coffee plant the produces low caffeine beans

2004 Korean scientists report human embryonic stem cell produced using a nucleus from an adult cell

2005 Korean scientists improve success rate of human adult nuclear transfer to embryonic cells by 10-fold

[COMMENT]—Portion of a protein file that gives a brief description of a protein.

[FEATURES]—Portion of a protein file that provides information about particular sections of a protein.

[FUNCTION]—Portion of a protein file that describe what the protein does.

[MISCELLANEOUS]—May give some specific added information about a protein within the protein file.

[SIMILARITY]—Portion of a protein file that indicates what family of similar proteins of which this one is a member.

[SUBCELLULAR LOCATION]—Portion of a protein file that indicates the kind of environment within the cell where this protein will likely be found.

[SUBUNIT]—Portion of a protein file that indicates whether or not this polypeptide is one of several that make up a protein.

3'—Refers to the end of DNA or RNA with an exposed hydroxyl group on the sugar's 3' carbon. The far end of each mRNA sequence is always the 3' end.

5'—Refers to the end of DNA or RNA with an exposed phosphate group. The beginning of each mRNA sequence is always the 5' end.

A—Adenine. One of the four bases found in DNA and RNA.

Accession number—Identifier for a file submitted to GenBank.

Adenosine di-phosphate—ADP. The uncharged chemical battery of the cell.

Adenosine tri-phosphate—ATP. The fully charged chemical battery of the cell.

ADP—adenosine di-phosphate, the uncharged chemical battery of the cell.

Alanine (A)—An amino acid.

Alpha helix—A cylindrical shape assumed by a portion of a protein that is stabilized by hydrogen bonds, a secondary protein structure.

Alternative splice sites—Areas where mRNAs can be cut more than one way to yield different proteins.

Amino acid—One of the 20 building blocks of proteins.

Amino hydrogen—The hydrogen bonded to the amino nitrogen of an amino acid.

Annotate—To record the specific properties, for example, the properties conferred by specific amino acids within a protein.

Apoptosis—Programmed cell death.

Anticodon—The three bases exposed on a transfer RNA that are complementary to three bases that make up a codon in mRNA.

Arginine (R)—One of the positively charged amino acids; the others are lysine (K) and histidine (H) (which is charged below pH 6).

Aspartate (D)—This is another name for the negatively charged amino acid, aspartic acid.

Aspartic acid (D)—This is another name a negatively charged amino acid, aspartate.

ATP—Adenosine triphosphate, the chemical energy packet or cell battery.

ATP synthase—A protein machine that runs on H^+ and makes ATP from ADP and phosphate.

Autosomes—Non-sex chromosomes; in humans the 22 pairs of chromosomes that are not the X and Y chromosomes.

BAC—*B*acterial *a*rtificial *c*hromosome that was used to clone large pieces of DNA of about 150,000 base pairs during the sequencing of genomes.

Backbone—The alternating sugar-phosphate sides of the DNA ladder or the N-C_α-C repeat structure of amino acids strung together to make a polypeptide. Because most RNA is single-stranded, RNA usually has only one backbone.

Base—One of the four substances that form the building blocks of DNA and code genetic information. DNA molecules are chains of four bases, adenosine (A), cytosine (C), guanine (G), and thymine (T), each slightly different chemically from the others.

Base pairs—The nucleotide A-T and G-C pairs that make up the rungs of the DNA ladder.

Beta sheet—One of the structural conformations of a protein where the primary amino acid sequence curves back upon itself to create a sheet of protein that looks like corrugated cardboard.

Bioinformatics—The computerized information files on genes, proteins and their functions and locations within cells. Bioinformatics includes the systems for organizing this information and the options that have been developed to search for particular information within this sea of data about biological systems.

C—Shorthand for the base cytosine in DNA or RNA.

Carbohydrate—Compound made of carbon, hydrogen and oxygen that forms an aldehyde or ketone with many hydroxyl groups. Sugars and starch are common carbohydrates.

Carbonyl oxygen—The oxygen attached to the last carbon in each amino acid.

Catalytic site—The position on an enzyme where reactants bind and are converted to the finished product.

cDNA—DNA made with reverse transcriptase and using mRNA as a template.

Centromere—Indentation on a chromosome where its identical sister chromatids are connected and where spindle fibers attach.

Chaperones—Special proteins that help other proteins fold into their precise three-dimensional shape.

Charged—Refers to amino acids that have either gained an electron or a proton and therefore carry either an extra negative or positive charge.

Chloroplast—The food factory of the cell that removes the carbon from carbon dioxide and uses it to make sugar.

Chromosome—A structure in the nucleus of a cell that contains one long molecule of DNA that contains the gene blueprint for the cell. One of each of our 23 pairs of chromosomes came from our father and one from our mother.

Circadian clock—The regular daily activities of a variety of organisms that are regulated by the sun. Light affects the cyclical production and degradation of particular proteins that alter the transcriptional pattern in the cell and the activity of the organism.

Clone—To capture a piece of DNA from one organism in the small genome of a bacterium or phage so as to isolate and amplify that piece of DNA.

Clone-based physical mapping—A technique used to determine a group of overlapping clones that together cover the length of entire chromosomes. Several identical pieces of DNA are cut into overlapping pieces of various sizes and cloned into bacteria. Then all of the clones are digested with one type of restriction enzyme and separated by size on a gel. Pieces that are the same size suggest that these clones might contain overlapping pieces of DNA. In this way, clones that are known to be derived from the same piece of DNA can be ordered to "cover" the sequence within that piece of DNA.

Codon—A precise series of three nucleotides in the mRNA that is complementary to the three nucleotide anticodon on tRNA. The codon determines which amino acid will be incorporated into a polypeptide.

Complementary strand—The opposite side of any DNA molecule, or the nucleic acid strand to which a single-stranded nucleic acid binds to. Because bases always pair in A-T, A-U, or G-C pairs, one side of a DNA molecule determines the sequence of its opposite side, which is the complementary strand.

Complexity—With regard to protein sequence, a sequence that does not use a lot of the same kinds of amino acids. Sequences that have many of the same kinds of amino acids lack complexity and are often excluded in searches for similar sequence because they tend to pull out other similar sequences that are not real matches.

Consensus sequence—The best common sequence derived from the sequences of a different proteins that contain the same type of structural domain. Mother Nature retains or conserves particular amino acids within portions of proteins that share a common structure and often function.

Conserved segment of synteny—A region along chromosomes in two different kinds of organisms that shares the same order of genes.

Contig—A continuous piece of DNA sequence created when BACs are arranged in sequence.

C-terminus—The far end of a protein. The "C" refers to the carbon in the COOH end of the last amino acid.

Curators—Individuals who point out the features of regions within genes and proteins cytoplasm and organize this information in databases.

Cytoplasm—The gelatin-like substance inside a cell.

DNA—Deoxyribonucleic acid, the molecules that transmit hereditary information.

Daughter cell—A cell formed as a result of mitosis.

Deoxyribonucleic acid—DNA, the molecule that transmits hereditary information.

Domain—A portion of a protein that has a particular structure and/or function.

E value—The "Expect value" that predicts the likelihood that an alignment of two sequences has occurred by chance in the absence of an evolutionary relationship. When the portions of alignment are identical, the Expect value is 0.0.

Electron—Negatively charged particle with almost no mass that occupies a defined space near the nucleus of an atom or more than one if the electron is shared.

Electron cloud—The shape that defines the space around the nucleus of an atom where electrons will be found.

Enzyme—Protein machine that catalyzes the production of a product from chemical reactants.

EST—See **Expressed sequence tag**.

Ether—A volatile organic molecule that can induce a dance in fruit flies that have a specific mutation called "ether a go-go" in a potassium ion channel. Also used as anesthesia.

Evolutionary tree—A branching diagram that indicates the evolutionary relationships among organisms.

Exons—The stretches of DNA that code for a protein.

ExPASy—Expert Protein Analysis System Proteomic Server at *http://www.expasy.org.*

Expect value—The "E value" that predicts the likelihood that an alignment of two sequences has occurred by chance in the absence of an evolutionary relationship. When the portions of alignment are identical, the Expect value is 0.0.

Expressed sequence tag (EST)—A sequence of a few hundred bases copied into DNA using the 3' end of an mRNA as a template.

Extracellular—Outside of a cell.

F_0—The motor of the ATP synthase embedded in the membrane that turns in response to H^+.

F_1—The portion of the ATP synthase that hangs outside of the membrane. This portion contains three bulbs that squeeze together ADP and phosphate to make ATP as the central asymmetrical gamma rod moves around and sequentially deforms the bulbs.

Family—A set of related protein structural domains.

FASTA format—A format that includes the gene or protein sequence in a format that can be used directly in a BLAST! Search.

Fat—Compounds made up of glycerol and three fatty acids.

Filter—Within the potassium channel, the narrowest portion that ensures that potassium and not sodium passes easily through the channel.

Functional domain—A portion of a protein that assumes a particular structure that has a particular job.

Functional group—A portion of a molecule such as an amino group (H_2N-) or carboxyl group (-COOH) that gives that portion of the molecule special properties.

G—Guanine, one of the four bases that makes up DNA or RNA.

Gamete—Egg and sperm cells.

Gap—A portion within an aligned sequence where there is no sequence match.

Gene—A section of a DNA molecule that codes for making a protein or family of proteins.

GeneChip®—A patented slide that can be engineered to contain an array of nucleic acids that correspond to a set of particular genes that can be screened simultaneously for binding by nucleic acids as a diagnostic test or analysis.

GeneOntology—A database that uses a limited set of words to allow one to easily find the molecular function, biological process involved, and location of a particular gene product. *http://www.geneontology.org/*.

Genetic analysis—Determining the function of a particular gene by altering a genome so the normal product is not made and then seeing the effect of this change on the organism.

Genome—The sum total of all the genes coded for by the haploid content of DNA, usually in the nucleus of an organism.

Genomics—The study of genomes, the entire haploid DNA content of an organism or organelle.

Glutamic acid (E)—A negatively charged amino acid.

Glycine—The smallest amino acid.

GYGD—The glycine, tyrosine, glycine, aspartic acid sequence found in the filter of most potassium channels.

Helix—The spiral shape assumed by the DNA ladder.

HGP—See **Human Genome Project**.

Hierarchical sequencing—An approach to sequencing a genome in which one first determines overlapping cloned segments of sequence and then determines the sequence within each of these overlapping pieces. The public consortium used this process to sequence the human genome.

High-throughput—Automated methods that allow multiple analyses to be conducted simultaneously.

Histidine (H)—An amino acid that is largely uncharged at pH 7, but positively charged at pH 6.

HomoloGene—A database within NCBI that allows one to find homologues among eukaryotic sequences from completed genomes. The graphical representation indicates conserved structural and functional domains within each protein.

Homologous—Similar.

Homotetramer—A protein such as the KcsA potassium channel made up of four identical subunits.

Human Genome Project (HGP)—The 13-year Human Genome Project to sequence the entire human genome that was funded and coordinated by the U.S. Department of Energy and National Institutes of Health in conjunction with the Wellcome Trust in England and partners in Japan, France, Germany, China, and others.

Hydrogen—The smallest element.

Hydrophilic (-ity)—Pertains to a molecule that "loves water." Refers to the property of some molecules, usually polar or charged molecules that do not hide from interacting and ordering the water around them.

Hydrophobic (-ity)—Pertains to a molecule that "fears water." Refers to the property of some molecules, nonpolar molecules that hide from interacting with and ordering the water molecules around them.

Identities—Within a sequence alignment, the percent of aligned amino acids that are identical.

IHGSC—See **International Human Genome Sequencing Consortium**.

Integral membrane protein—A protein found within the plasma membrane.

International Human Genome Sequencing Consortium—Group of laboratories throughout the world responsible for the effort to sequence the human genome that resulted in the free deposition of the entire sequence in the public databases.

Intracellular—Inside a cell.

Introns—The portions of an mRNA that are spliced out during mRNA processing, before the mRNA exits the nucleus. Introns may contain sequences that regulate transcription of DNA to mRNA.

Ion channel—A protein within a membrane that allows for the regulated passage of ions from one side of the membrane to the other.

Lysine (K)—A positively charged amino acid.

Map Viewer—Program accessible from the NCBI database "Gene" page that allows one to examine conserved regions of synteny along homologous human, mouse, and rat chromosomal regions.

Mass spectrometry—A process that allows one to determine the proteins present in a sample.

Meiosis—The process whereby a cell reduces its genetic information by half during the production of eggs and sperm.

Messenger RNA (mRNA)—Single-stranded nucleic acid made using the DNA for one gene as a template. After processing to remove introns in eukaryotes, the processed mRNA is sent out of the nucleus to the ribosome to code for making a protein.

Microarray—A regular pattern of different nucleic acids or proteins arranged in rows on a slide. Microarrays are used to screen an entire genome for global mRNA production or the binding of some other molecule.

Microbes—Simple one-celled organisms such as bacteria and archaea.

Mitochondrion—Organelle where the chemical battery of the cell, ATP, is made.

Mitosis—Eukaryotic cell division.

Model organism—Organism such as phage, bacteria, yeast, the weed *A. thaliana*, fruit flies, *C. elegans*, zebrafish, rat, or mouse that has a relatively short life cycle that is genetically manipulated to learn the effects of altering specific genes.

Molecular genetics—Genetic analysis at the level of protein-protein or protein-nucleic acid interactions.

mRNA—See **Messenger RNA**.

Mutation—An alteration in the DNA sequence of a gene.

Nonredundant (Nr)—Database that has been stripped of multiple files or records for the same gene or protein.

Nr—See **Nonredundant.**

N-terminus—The beginning of a protein. This is the end that has an exposed amino functional group.

Nuclear membrane—Membrane around the nucleus that also serves as part of the endoplasmic reticulum.

Nucleus—The membrane-bound bag-like structure within a cell that contains the cell's DNA.

Open reading frame (ORF)—The expanse of DNA that codes for a particular protein.

Operons—A group of genes controlled by a specific regulatory region.

ORF—See **Open reading frame**.

Organelle—A specialized region within a cell.

Organism—Any living thing.

Oxygen—Element used by humans and many other organisms for respiration, to carry out oxidative phosphorylation after breaking down food.

P arm—The short arm of a eukaryotic chromosome.

PCR—See **Polymerase chain reaction.**

Phosphate group (P_i)—PO_4 group that when added to ADP creates ATP. This functional group is also known as inorganic phosphate.

P_i—Inorganic phosphate (PO_4) group that when added to ADP creates ATP.

Plasma membrane—The membrane made of phospholipids and embedded and attached proteins that surrounds the outside of the cell. The plasma membrane and the proteins embedded within regulate the flow of materials in and out of the cell.

Polarity—Describing that property of having a partial or full positive or negative charge.

Poly-A tail—A string of A's that are added to the end of mRNA.

Polymerase chain reaction (PCR)—Method using a heat-resistant DNA synthesizing enzyme and primers that flank the section of DNA to be amplified to make many copies of one section of DNA that includes the primers.

Polypeptide—A generic name for a string of amino acids. This name comes from the fact that amino acids are connected to each other by a peptide bond. Some protein machines are made up of only one polypeptide, while others are made up of several polypeptides folded together.

Pore helices—These are the four alpha helices within the potassium channel that are positioned so as to attract the incoming potassium ion. Note that each pore helix, like all alpha helices, has polarity and the end with a partial negative charge is aimed toward the positively charged potassium ion.

Positively charged ion—An ion that has either lost an electron or gained a proton. The side chains of the amino acids arginine and lysine each acquire a proton at the normal pH 7.4 of the cell.

Positives—Refers to the percent of similarity within the portion of two sequences that demonstrates alignment. Positives include not only amino acids that are identical but amino acids for which Mother Nature has allowed a conservative change to an amino acid that has similar properties.

Potassium ion—A positively charged atom that is normally kept at a relatively high concentration within the cell by potassium pumps. This ion is released when a potassium channel opens. Potassium channels are opened during the transmission of a neural impulse to return the cell to its normal negatively charged state. These channels respond to the opening of adjacent sodium channels that allowed an influx of sodium ion. These alternating fluxes of ions move along the axon (the long, narrow portion) of a neural cell and constitute a nerve impulse.

Probe hybridization—A process by which bacterial cells are tagged with a probe that allows researchers to see if the desired STS sequence is present.

Processed mRNA—mRNA that has been spliced to remove sequences called introns.

Protein—A large molecule made up of one or more chains of amino acids.

Protein domain—A portion of a protein that has a particular structure and associated function. The same protein domain may appear within proteins that have very different kinds of jobs within the cell. One mechanism by which new proteins evolve is through shuffling of protein domains.

Protein query vs. translated database—This BLAST option called tblastn searches GenBank for sequences similar to a protein sequence by asking the computer to search for a possible nucleotide sequence that, when translated, would produce a protein match.

Proteomics—The study of all of the proteins produced by a particular type of organism. This study may include the prediction of protein products based upon DNA sequence; the separation and identification of proteins within parts of the cell, or within specific tissues or by the whole organism; and the determination and analysis of the structures of particular proteins.

Proton—A positively charged particle with a relative weight of about 1 in the nucleus of an atom. The number of protons in the nucleus determines which kind of element it is.

Proton pump—An ATP synthase in reverse that pumps protons into the cell.

Pseudogenes—A gene that is not active.

PUBMED—A database within the NCBI suite of databases that allows one to access abstracts or summaries of scientific journal articles about particular topics. Links are also provided to the full text for a price.

Q arm—The long arm of a human chromosome.

Radiation hybrid mapping—A process in which human chromosomes are cut by irradiation so they can fuse with hamster DNA inside many hamster cells. They are then tagged so that scientists can determine which genes or STS sequences appear near each other on human chromosomes. In this way, a physical map of the approximate order of certain genes and STS sequences along the human chromosomes was determined.

Random shotgun sequencing—Determining the A, C, T, G sequence of cloned DNA without first attempting to order the clones in any way.

Replicate—Make an exact copy of DNA.

Residue—Another name for amino acid. Amino acids within a protein or polypeptide sequence are often referred to as residues.

Restriction enzyme—An enzyme that cuts DNA at a region defined by a sequence that is usually 4 to 6 base pairs long.

Ribonucleic acid (RNA)—A single-stranded nucleic acid made up of a chain of nucleotides and containing a ribose sugar component. It is involved in gene expression.

RNA—See **Ribonucleic acid.**

RNA polymerase—The enzyme that controls RNA synthesis.

Secondary structure—The precise folds of portions of polypeptides. Two examples of secondary structure are the alpha helix and the beta sheet.

Selectivity filter—The region of an ion channel that has been conserved by Mother Nature because it restricts access in and out of the channel to a particular type of ion.

Sequence/alignment viewer—This tool within the program Cn3D allows one to correlate a position on the structure of a protein domain with a particular amino acid within the sequence for that structure. Since the amino acid sequences of several different proteins having this same structural domain are aligned with the sequence of the structure being examined, one may also view regions that remain invariant (have been conserved) among the proteins listed.

Sequence motif—A pattern found in the amino acid sequence of a protein that predicts the particular use of that section of the protein, such as a site that is likely to be phosphorylated. Some sequence motifs are invariant; others give options for particular residues within a string of residues.

Sequence tagged sites—Abbreviated STSs. These are physical regions along the chromosome that were determined by a variety of means to have a particular sequence before the entire chromosome was sequenced.

Sex chromosomes—Chromosomes that carry the genes that determine the sexual characteristics of a living thing. In humans, the sex chromosomes are the X and the Y chromosomes.

Side chain—The part of the amino acid that distinguishes one amino acid from another. The side chain is always attached to the alpha carbon in the N-C_α-C backbone of the amino acid.

Single nucleotide polymorphism—Abbreviated SNP, this variation is usually a single base change that creates an alternative DNA sequence for a particular gene. Some SNPs appear to have no effect, but others are associated with disorders that can be inherited or arise through spontaneous modification of DNA.

SNP—See **Single nucleotide polymorphism**.

Sodium ion—Positively charged atom of the element sodium that has lost one of its electrons. This ion passes through a potassium channel inefficiently, even though it is smaller than a potassium ion.

Sporulation—The process induced by nitrogen-starvation whereby a diploid yeast cell undergoes a transformation into four haploid spores.

Structural domain—A portion of a protein structure that consistently assumes a particular shape among different proteins. Often this portion of the protein carries out a function similar to its role within other proteins.

STSs—See **Sequence tagged sites**.

Subunits—Can refer to bases within DNA or RNA or to amino acids within a polypeptide.

Summary—A display option for an NCBI Gene page that always contains particular kinds of information about that gene such as alternative names or aliases and the chromosomal location.

Sv—Link from Map Viewer to Sequence Viewer with regard to a particular gene.

T—The DNA base called thymine. Thymine always pairs with adenine to form a rung of the DNA ladder.

Tblastn—Protein query vs. translated database. This BLAST option searches GenBank for sequences similar to a protein sequence by asking the computer to search for a possible nucleotide sequence that, when translated, would produce a protein match.

Template—A pattern. One side of an unzipped DNA molecule can provide the pattern or template for building the other half of the DNA molecule. This property arises because of the constancy of the A-T and G-C pairing rules.

Tissues—A specialized mass of cells, such as muscle.

Top-down sequencing—An approach to sequencing a genome in which one first determines overlapping cloned segments of sequence and then determines the sequence within each of these overlapping pieces. The public consortium used this process to sequence the human genome.

Transcribe—To make mRNA using the DNA of a gene as a template. The cell carries out this process in the nucleus using the four nucleoside triphosphates as raw materials.

Transfer RNA (tRNA)—RNA that shuttles an amino acid to the site of protein production, the ribosome.

Translate—To convert the mRNA code into a protein. The cell carries out this function at the ribosome using the mRNA code and tRNAs that bring particular amino acids to be linked to make a protein.

Transmembrane—Crossing the membrane of a cell.

Transposon—A DNA sequence that codes for enzymes needed to either make a copy of this sequence and insert that copy into another region in the genome or to cut out this sequence and insert it into another region. Some transposons have lost the ability to make these enzymes but they nonetheless can be recognized and duplicated or moved by enzymes made by intact transposons.

tRNA—See **Transfer RNA**.

Tumor suppressor—A gene that normally puts the brakes on cell division. BRCA1 and P53 are example of tumor suppressors that when mutated become oncogenes that allow uncontrolled cell division or cancer.

Tyrosine (Y)—An amino acid that has a ring structure and polar hydroxyl group.

UnitProt/Swiss-Prot—UnitProt is the most comprehensive worldwide database of protein information. It is made up of the Swiss-Prot annotated database of known proteins, the TrEMBL database of proteins predicted based upon DNA sequence, and another protein database called PIR.

Untranslated Region—The untranslated portion of a processed mRNA.

UTR—See **Untranslated Region**.

Whole shotgun sequencing—Determining the sequence of the ends of individual cloned pieces of DNA and then using computer power to arrange the pieces in the order of a genome.

X-ray crystallography—A method for determining the location of each atom within the structure of a protein. When x-rays are directed at a crystal of the protein, the pattern they create can be analyzed mathematically to give these locations.

CHAPTER 1

1. Stephen Dickman, "Turning on and turning off the immune system," *Exploring the Biomedical Revolution*, ed. Maya Pines, Baltimore, MD and London: Howard Hughes Medical Institute, 1999, p. 341.
2. Michel Roger, "Influence of host genes on HIV-1 disease progression," *FASEB J* 12 (1999): pp. 625–632.
3. J. D.Watson and F.H.C. Crick, "Molecular Nature of Nucleic Acids," *Nature* 171 (1953): p. 737.

CHAPTER 2

1. Eric Gouaux, "Single potassium ion seeks open channel for transmembrane travels: tales from the KcsA structure," *Structure* 6 (1998): 1221–1226.
2. Declan Doyle, et al., "The Structure of the Potassium Channel: Molecular Basis of K^+ Conduction and Selectivity," *Science* 280 (1998): 69–77.
3. Christina I. Petersen, et al., "*In vivo* identification of genes that modify ether-a-go-go-related gene activity in *Caenorhabditis elegans* may also affect human cardiac arrhythmia," *Proceedings of the National Academy of Sciences USA* 101 (2004): 11773-11778.

CHAPTER 3

1. International Human Genome Sequencing Consortium, "Initial sequencing and analysis of the Human Genome," *Nature* 409 (2001): 860–920.
2. *Ibid.*
3. J. Craig Venter, et al., "The sequence of the human genome," *Science* 291 (2001): 1304–1351.
4. International Human Genome Sequencing Consortium, *op. cit.*, p. 862.
5. J.C. Venter, et al., *op. cit.* This value for the length of the coding region is calculated from the percent of base pairs spanned by exons relative to that spanned by introns and average gene size given.
6. International Human Genome Sequencing Consortium, *op. cit.*
7. X. She, et al., "Shotgun sequence assembly and recent segmental duplications within the human genome," *Nature* 431 (2004): 927–930.
8. Greg Gibson and Spencer V. Muse, *A Primer of Genome Science.* Sunderland, Massachusetts, Sinauer Associates, 2002.
9. J. Craig Venter, *op.cit.*
10. International Human Genome Sequencing Consortium, *op. cit.*
11. David Baltimore, "Our genome revealed," *Nature* 409 (2001): p. 814.

CHAPTER 4

1. David Baltimore, "Our genome unveiled," *Nature* 409 (2001): 814–816.

2. International Human Genome Sequencing Consortium, "Initial sequencing and analysis of the Human Genome," *Nature* 409 (2001): 860–920.
3. J. Craig Venter, et al., "The Sequence of the Human Genome," *Science* 291 (2001): 1304–1351.
4. *Ibid.*
5. J. Craig Venter, et al., *op. cit.*
6. X. She, et al., "Shotgun sequence assembly and recent segmental duplications within the human genome," *Nature* 431 (2004): 927–930.
7. Lincoln Stein, "Human Genome: end of the beginning," *Nature* 431 (2004): 915–916.
8. IHGSC, "Finishing the euchromatic sequence of the human genome," *Nature* 431 (2004): 931–945.
9. IHGSC (2001), *op. cit.*
10. Wen-Hsiung Li, Zhenglong Gu, Haidong Wang, and Anton Nekrutenko, "Evolutionary analyses of the human genome," *Nature* 409 (2001): 847–849.
11. J. Craig Venter, *op. cit.*
12. IHGSC (2001) *op. cit.*, p. 906.
13. National Institutes of Health press release, "Researchers compare chicken, human genomes," Dec. 8, 2004 at *www.genome.gov/12514316.*
14. International Chicken Genome Sequencing Consortium, "Sequence and comparative analysis of the chicken genome provide unique perspectives on vertebrate evolution," *Nature* 432 (2004): 695.
15. Olivier Jaillon, et al., "Genome duplication in the teleost fish *Tetraodon nigroviridis* reveals the early vertebrate proto-karyotype," *Nature* 431 (2004): 946–957.
16. International Chicken Genome Sequencing Consortium, *op. cit.*
17. Olivier Jaillon, et al., *op. cit.*
18. Rat Genome Sequencing Project Consortium, "Genome sequence of the Brown Norway rat yields insights into mammalian evolution," *Nature* 428 (2004): 493.
19. *Ibid.*
20. *Ibid.*
21. Andrew G. Clark, "Inferring non-neutral evolution from human-chimp-mouse othologous gene trios," *Science* 302 (2003): 1960–1963.
22. Maynard V. Olson and Ajit Varki, "Sequencing the chimpanzee genome: insights into human evolution and disease," *Nature Reviews: Genetics* 4 (2002): 20–28.
23. Andrew G. Clark, *op. cit.*
24. S. Naz, et al., "Distinctive audiometric profile associated with DFNB21 alleles of TECTA," *Journal of Medical Genetics* 40 (2003): 360–363.

25. W. Enard, et al., "Molecular evolution of *FOXP2*, a gene involved in speech and language," *Nature* 418 (2002): 869–872.

26. Elizabeth Pennisi, "Genome comparisons hold clues to human evolution," *Nature*: 302 (2003): 1876–1877.

27. Maynard V. Olson and Ajit Varki, *op. cit.*

CHAPTER 5

1. Jessica Queller, "Cancer and the Maiden," *New York Times*, March 5, 2005, p. A27.

2. E. Lander, "Scanning Life's Matrix: Genes, Proteins and Small Molecules" Howard Hughes Medical Institute Holiday Lecture 2002, HHMI, Maryland.

3. Eric N. Shiozaki, et al., "Structure of the BRCT repeats of BRCA1 bound to a BACH1 phosphopeptide: implications for signaling," *Molecular Cell* 14 (2004): 405–412.

4. R. Scott Williams, et al., "Structural basis of phosphopeptide recognition by the BRCT domain of BRCA1" *Nature Structural and Molecular Biology* 11(2004): 519–525.

5. Julie A. Clapperton, et al.,"Structure and mechanism of BRCA1 BRCT domain recognition of phosphorylated BACH1 with implications for cancer" *Nature Structural and Molecular Biology* 11 (2004): 512–518.

6. Eric Martz, "Protein Explorer: Easy Yet Powerful Macromolecular Visualization," *Trends in Biochemical Sciences* 27 (2002): 107–109. *http://proteinexplorer.org*

7. Annachiarra De Sander-Giovannoli, et al., "Lamin A truncation in Hutchison-Gilford Progeria," *Science* 300 (2003): 2055.

8. Eriksson, et al., "Recurrent de novo point mutations in lamin A cause Hutchinson-Gilford progeria syndrome," *Nature* 423 (2003): 293–298.

9. Francis Collins, NIH News Release at *http://www.genome.gov/11006962* 2003.

10. C. Holden, "Schizophrenia in the fast lane," *Science* 301 (2003): 1645.

11. C. Holden, "Deconstructing Schizophrenia," *Science* 299 (2003): 333–335

12. S. Chu, et al., "The transcription program of sporulation in budding yeast," *Science* 282 (1998): 699–705.

13. C. M. Perou, et al., "Molecular portraits of human breast tumours," *Nature* 406 (2000): 747–752.

14. Schindler, et al., "Structural mechanism for STI-571 inhibition of Abelson tyrosine kinase" *Science* 289 (2000): 1938–1942.

15. Eric Lander, *op. cit.*

CHAPTER 6

1. Declan Doyle, et al., "The Structure of the Potassium Channel: Molecular Basis of K^+ Conduction and Selectivity," *Science* 280 (1998): 69–77.
2. Declan Doyle, et al., *op. cit.*
3. S. Altschul, et al., "Basic Local Alignment Search Tool," *J Mol Biol* 215 (1990): 403–410.
4. K. Dedek, et al., "Myokymia and neonatal epilepsy caused by a mutation in the voltage sensor of the KCNQ2 K^+ channel," *Proceedings of the National Academy of Sciences USA* 98 (2001): 12272–12277.
5. Youxing Jiang, et al., "X-ray structure of a voltage-dependent K+ channel," *Nature* 423 (2003): 33–41.

CHAPTER 7

1. Ray Tenebruso, personal communication.
2. P.D. Boyer, "The binding change mechanism for ATP synthase—Some probabilities and possibilities," *Biochimica et Biophysica Acta* 1140 (1993): 215–250.
3. J. P. Abrahams, A. G. Leslie, R. Lutter, and J. E. Walker, "Structure at 2.8 Å resolution of F_1-ATPase from bovine heart mitochondria," *Nature* 370 (1994): 621–628.
4. M. Kanehisa, "Toward pathway engineering: a new database of genetic and molecular pathways," *Science & Technology Japan* 59 (1996): pp. 34–38.
5. *Ibid.*

BOOKS AND ARTICLES

Abrahams, J. P., et al. "Structure at 2.8 Å resolution of F1-ATPase from bovine heart mitochondria." *Nature* (370) 1994: 621–628.

Alizadeh, A. A., et al. "Distinct types of diffuse large B-cell lymphoma identified by gene expression profiling." *Nature* (403) 2000: 503–511.

Baltimore, David. "Our genome revealed." *Nature* (409) 2001: 814–816.

Bell, J. "Predicting disease using genomics." *Nature* (429) 2004: 453–456.

Boyer, Paul D. "The binding change mechanism for ATP synthase–Some probabilities and possibilities." *Biochemica and Biophysica Acta—Bioenergetics* (1140) 1993: 215–250.

Campbell, A. Malcolm, and Laurie J. Heyer. *Discovering Genomics, Proteomics and Bioinformatics.* San Francisco: Pearson Education, Inc., Benjamin Cummings, 2003.

Chu, S., et al. "The transcription program of sporulation in budding yeast." *Science* (282) 1998: 699–705.

Clapperton, Julie A., Isaac A. Manke, Drew M. Lowery, et al. "Structure and mechanism of BRCA1 BRCT domain recognition of phosphorylated BACH1 with implications for cancer." *Nature Structural and Molecular Biology* (11) 2004: 512–518.

Clark, Andrew G., et al. "Inferring nonneutral evolution from human-chimp-mouse othologous gene trios." *Science* (302) 2003: 1960–1963.

Claverie, Jean-Michel, and Cedric Notredame. *Bioinformatics for Dummies.* New York: Wiley Publishing, Inc., 2003.

Cloninger, R. "The discovery of susceptibility genes for mental disorders." *Proceedings of the National Academy of Sciences* (99) 2002: 13365–13367.

Dedek, K., et al. "Myokymia and neonatal epilepsy caused by a mutation in the voltage sensor of the KCNQ2 K+ channel." *Proceedings of the National Academy of Sciences USA* (98) 2001: 12272–12277.

De-Sander-Giovannoli, Annachiarra, et al. "Lamin A truncation in Hutchinson-Gilford Progeria." *Science* (300) 2003: 2055.

Enard, W., et al. "Molecular evolution of FOXP2, a gene involved in speech and language." *Nature* (418) 2002: 869–872.

Eriksson M., et al. "Recurrent de novo point mutations in lamin A cause Hutchinson-Gilford progeria syndrome." *Nature* (423) 2003: 293–298.

Gibson, Greg, and Spencer V. Muse. *A Primer of Genome Science.* Sunderland, Massachusetts: Sinauer Associates, 2002.

Gouaux, Eric. "Single potassium ion seeks open channel for transmembrane travels: tales from the KcsA structure." *Structure* (6) 2003: 1221–1226.

Hartl, Daniel L., and Elizabeth W. Jones. *Genetics, Analysis of Genes and Genomes*, 6th ed. Sudbury, Massachusetts: Jones and Bartlett, 2005.

Holden, C. "Deconstructing Schizophrenia." *Science* (299) 2003: 333–335.

Holden, C. "Schizophrenia in the fast lane." *Science* (301) 2003: 1645.

International Human Genome Sequencing Consortium. "Finishing the euchromatic sequence of the human genome." *Nature* (431) 2004: 931–945.

International Human Genome Sequencing Consortium. "Initial sequencing and analysis of the human genome." *Nature* (409) 2001: 860–920.

Jaillon, Olivier, et al. "Genome duplication in the teleost fish *Tetraodon nigroviridis* reveals the early vertebrate proto-karyotype." *Nature* (431) 2004: 946–957.

Jiang, Youxing, et al. "X-ray structure of a voltage-dependent K^+ channel." *Nature* (423) 2003: 33–41.

Kyte, J., and R. Doolittle. "A simple method for displaying the hydropathic character of a protein." *Journal of Molecular Biology* (157) 1982: 105–132.

Martz, Eric. "Protein Explorer: easy yet powerful macromolecular visualization." *Trends in Biochemical Sciences* (27) 2002: 107–109.

Mullis, Kary B. *Dancing Naked in the Mind Field.* New York: Pantheon Books, 1998.

Naz, S., et al. "Distinctive audiometric profile associated with DFNB21 alleles of TECTA." *Journal of Medical Genetics* (40) 2003: 360–363.

Olshevsky, George, ed. *Understanding the Genome.* New York: Scientific American, Byron Preiss Visual Publication, Inc., Warner Books, 2002.

Olson Maynard V., and Ajit Varki. "Sequencing the chimpanzee genome: insights into human evolution and disease." *Nature Reviews: Genetics* (4) 2003: 20–28.

Pennisi, Elizabeth. "Genome comparisons hold clues to human evolution." *Science* (in News of the Week) (302) 2003: 1876–1877.

Perou, C.M., et al. "Molecular portraits of human breast tumours." *Nature* (406) 2000: 747–752.

Pines, Maya, editor. *Exploring the Biomedical Revolution.* Bethesda, Maryland: Howard Hughes Medical Institute, 1999.

Queller, Jessica. "Cancer and the Maiden." *New York Times*, March 5, 2005: p. A27.

Rat Genome Sequencing Project Consortium. "Genome sequence of the Brown Norway rat yields insights into mammalian evolution." *Nature* (428) 2004: 493–521.

Ridley, Matt. *Genome, the Autobiography of a Species in 23 Chapters.* New York: Harper Collins, 1999.

Roger, Michael. "Influence of host genes on HIV-1 disease progression." *FASEB Journal* (12) 1999: 625–632.

Schindler, et al. "Structural mechanism for STI-571 inhibition of Abelson tyrosine kinase." *Science* (289) 2000: 1938–1942.

She, X., et al. "Shotgun sequence assembly and recent segmental duplications within the human genome." *Nature* (431) 2004: 927–930.

Shiozaki, Eric N., Lichuan Gu, Nieng Yan, and Yigong Shi. "Structure of the BRCT repeats of BRCA1 bound to a BACH1 phosphopeptide: implications for signaling." *Molecular Cell* (14) 2004: 405–412.

Simon, J., and A. Bedalov. "Perspectives: Yeast as a model system for anticancer drug discovery." *Nature Reviews-Cancer* (4) 2004: 1–8.

Stein, Lincoln D. "Human genome: end of the beginning." *Nature* (431) 2004: 915–916.

Venter, J. Craig, et al. "The Sequence of the human genome." *Science* (291) 2001: 1304–1351.

Watson, James D., and Francis H. C. Crick. "Molecular Nature of Nucleic Acids." *Nature* (171) 1953: 737.

Wen-Hsiung, Li, Zhenglong Gu, Haidong Wang, and Anton Nekrutenko. "Evolutionary analyses of the human genome." *Nature* (409) 2001: 847–849.

Werth, Barry. *The Billion Dollar Molecule, One Company's Quest for the Perfect Drug.* New York: Simon and Schuster, 1994.

Williams, R. Scott, Megan S. Lee, D. Duong Hau, and J. N. Mark Glover. "Structural basis of phosphopeptide recognition by the BRCT domain of BRCA1." *Nature Structural and Molecular Biology* (11) 2004: 519–525.

Biomolecular Interaction Network Database (BIND)
http://bind.ca/

Dolan DNA Learning Center of Cold Spring Harbor Laboratory
http://www.dnai.org/

Ensembl!
http://www.ensembl.org

ExPASy (Expert Protein Analysis System) Proteomics Server
http://kr.expasy.org/

FlyBase
http://flybase.net

Howard Hughes Medical Institute
http://www.hhmi.org/biointeractive/

Kyoto Encyclopedia of Genes and Genomes
http://www.genome.jp/kegg/

National Center for Biotechnology Information (NCBI)
http://www.ncbi.nlm,nih.gov/

NCBI Blast!
http://www.ncbi.nlm.nih.gov/BLAST/

NCBI Online Mendelian Inheritance in Man
http://www.ncbi.nlm.nih.gov/entrez/query.fcgi?db=OMIM

Nobel Prize Winners
http://nobelprize.org/

Protein Data Bank
http://www.rcsb.org/pdb/

Protein Explorer FrontDoor
http://molvis.sdsc.edu/protexpl/frntdoor.htm

UnitProt/Swiss-Prot
http://www.ebi.ac.uk/swissprot/

U.S. National Institutes of Health
http://www.nih.gov/

WormBase
http://www.wormbase.org

page:

4: © 21st Century Publishing
6: © Peter Lamb
9: © Peter Lamb
11: © Peter Lamb
16: (top) © Peter Lamb
16: (bottom) Courtesy Jean-Yves Sgro, the University of Wisconsin, Madison. Adapted from Doyle, et al., 1998.
21: (top) © Peter Lamb
21: (bottom) © 21st Century Publishing
24: © Peter Lamb
26: © Peter Lamb
28: © Peter Lamb
42: © Peter Lamb
45: © 21st Century Publishing
46: © Jeffrey L. Rotman/CORBIS
51: © Peter Lamb
54: © Peter Lamb
59: © Peter Lamb
70: © 21st Century Publishing
75: © Peter Lamb
88: © 21st Century Publishing
90: © Peter Lamb
97: © Peter Lamb
106: © Peter Lamb
110: Ann Finney Batiza
111: Ann Finney Batiza
112: Courtesy NCBI
119: © Peter Lamb
121: © Peter Lamb
131: © Peter Lamb
134: Ann Finney Batiza
135: Courtesy NCBI
139: (top) Courtesy Jean-Yves Sgro, the University of Wisconsin, Madison
139: (bottom) Courtesy NCBI
144: Ann Finney Batiza
145: Courtesy NCBI

ANN FINNEY BATIZA grew up in Texas and Oklahoma. She was interested in science from her earliest years and developed a lifelong interest in biology under the tutelage of her high school biology teacher, Mrs. Lois Brandt. At the urging of her cousin, Dr. Batiza attended Lawrence University in Appleton, Wisconsin, where she received a B.A. in biology and chemistry, with emphasis in biology, in 1969. She then went on to the Department of Microbiology at Yale University, but left to move to California, and earned an M.A. in education from San Diego State University. After moving to St. Louis, she developed the educational content of the new Discovery Room at the St. Louis Science Center. She then taught high school and wrote science education materials for several years before moving with her two children to Madison, Wisconsin. She earned a Ph.D. in cellular and molecular biology from the University of Wisconsin in 1998 in the laboratory of Professor Patrick Masson. For her postdoctoral work and later as a researcher, she continued to study mechanosensitive ion channels in the U.W. laboratory of Professor Ching Kung.

Dr. Batiza has lectured at Washington University in St. Louis (in education), at Madison Area Technical College, and at the University of Wisconsin, Madison, and is the author of scientific papers in the fields of bacterial UV sensitivity, yeast calcium physiology, and bacterial and yeast ion channels. She was also a delegate to a National Political Convention in 2004.